Seeds for Diversity and Inclusion

Yoshiaki Nishikawa · Michel Pimbert
Editors

Seeds for Diversity and Inclusion

Agroecology and Endogenous Development

Editors
Yoshiaki Nishikawa
Faculty of Economics
Ryukoku University
Kyoto, Japan

Michel Pimbert
CAWR
Coventry University
Coventry, UK

ISBN 978-3-030-89404-7   ISBN 978-3-030-89405-4   (eBook)
https://doi.org/10.1007/978-3-030-89405-4

Cover illustration: © Melisa Hasan

This Palgrave Macmillan imprint is published by the registered company Springer Nature Switzerland AG
The registered company address is: Gewerbestrasse 11, 6330 Cham, Switzerland

FOREWORD

How can we feed the world sustainably? This is a perennial issue, yet one that has never been high on the global agenda until now—in the Anthropocene, a new epoch marked by the impact of humanity on planetary systems.

The means to this end, however, reveal sharp differences. On the one hand, there are the hyper-technological solutions offered by modern science. On the other are approaches that respect local and Indigenous knowledge, harmonized with ecological conditions and historically demonstrated by the survival of agrarian societies.

Seeds constitute one of the prime subjects in this existential debate. Fashioned through a long collaboration between humankind and nature, seed is a semi-artifact. The saving and cultivation of seeds are practices deeply embedded in the peasant way of life, ranging from their material use in farming and daily diet to their spiritual aspect, interconnecting the human–nature relationship.

Seeds for Diversity and Inclusion explores seeds across different cultures, with a major focus on Asian countries and areas, and within a range of interrelated contexts, from agroecology and sovereignty to endogenous development—determined by local values, efforts and benefits. Ecological, ontological and institutional aspects of seeds in Asia and other regions help us understand the realities of rural societies in a globalizing world. By investigating the work of smallholders, seed sharers, traditional family-led companies and other actors in this arena, we can see many potentials through which crop diversity and food security can be sustained in a climate-changed world.

Kyoto, Japan   Motoki Akitsu
August 2021

PREFACE

The inspiration for this book came in 2009, when the first editor, Yoshiaki Nishikawa, came across a book chapter written by the second editor, Michel Pimbert. This was 'Transforming knowledge and ways of knowing', published in *Towards Food Sovereignty: Reclaiming Autonomous Food Systems* (IIED). Nishikawa was trying to engage with the concept of food sovereignty, which was being actively theorized in Western academic circles, and finding some difficulty in interpreting and applying the idea in the East Asian context.

Many academics and practitioners now recognize the importance of different forms of knowledge, especially as it relates to traditional knowledge for the realization of a more sustainable society. However, the processes by which such forms of knowledge are evaluated should be of critical importance. Often, knowledge is interpreted through the lens of norms established within Western academia. Knowledge, however, is embedded in ways of knowing, and those must be understood as a relational process within a particular, often very local, context.

Seeds for Diversity and Inclusion aims to contribute to a more nuanced debate around the many approaches to seed saving and cultivation that go beyond the dominant dichotomous conceptualization of seed governance, often characterized as traditional vs modern, subsistence vs commercial or local vs global. While reflecting on the increasing concentration of power among a handful of corporations in the current seed system, we argue that such classifications limit our ability to critically reflect on and acknowledge humanity's diverse relationships with seeds around the world, which create a mosaic of dynamic complementarities and autonomous practices that shape the ways we understand our societies.

We want to invite readers to join us in the adventure of questioning our ways of knowing, in part by engaging with a diversity of case studies on 'all things seed' (cultivation, saving, trading, sovereignty and more) from East Asia to Bhutan, Scotland and beyond.

The chapters in this book are the outcome of research from a number of projects funded by different organizations. However, the book itself is a product of three converging networks. The first is the Centre for Agroecology, Water and Resilience at Coventry University, UK, and its partners. The second is a network established around a project on the sustainability of seed procurement by small-holder farmers in Asia (Project # 17H04627), funded by the Japan Society for Promotion of Sciences. And the third is a network spinoff of the FEAST project (formally titled Lifeworlds of Sustainable Food Consumption and Production: Agri-food Systems in Transition) at the Research Institute for Humanity and Nature in Kyoto, Japan.

This confluence was made possible through the visiting researcher programme of Ryukoku University in Kyoto, which hosted a study—'Discourse study on management of biodiversity for food and agriculture'. This visiting researcher programme allowed Yoshiaki Nishikawa to be based at Coventry University where we joined forces to make this book a reality.

Finally, we would like to thank Barbara Kiser for her diligent and thoughtful review and careful editing of our initial manuscript. We also thank the team at Palgrave Macmillan—especially Abarna Antonyraj and Rachael Ballard.

Kyoto, Japan   Yoshiaki Nishikawa
Coventry, UK   Michel Pimbert

CONTENTS

NOTES ON CONTRIBUTORS

Alejandro Argumedo is Director of Programs and Andean Amazon Lead of Swift Foundation.

Stan Blackley is the Programme Leader for the M.Sc. Gastronomy programme at Edinburgh's Queen Margaret University, which takes a multidisciplinary approach to examining the many ways that food shapes society and the world around us.

Salvatore Ceccarelli has been full professor of Agricultural Genetics at the Institute of Plant Breeding, University of Perugia. His areas of expertise include participatory and evolutionary plant breeding as well as relationships between biodiversity, food, health and climate change.

Stefania Grando is a plant breeder with long-term experience in the management and leadership of agricultural research.

Makoto Kawase and Kenji Irie work for Tokyo University of Agriculture, Japan.

Ayako Kawai completed her Ph.D. in Human Ecology at Australian National University. Interest in human–nature relationships, bio-cultural diversity, sustainable seed system and culturally specific understandings and valuing of plants and other natural resources, based in Tokyo, Japan.

Mai Kobayashi is currently a researcher at the Kyoto University Asian Studies Unit and the Graduate School of Economics. Her interests are in processes and expressions of de- and re-peasantization in a post development context.

David McVey is a restaurateur and community food researcher in Edinburgh, Scotland.

Mami Nagashima worked as a coordinator of research and development project for plant resources in Myanmar, implemented by Kochi Prefectural Makino Botanical Garden, Japan.

Yoshiaki Nishikawa is Professor of Agricultural Economics and Resource Management in the Faculty of Economics, Ryukoku University, established in 1639 based on the spirit of the true teaching of Pure Land Buddhism (Jodo Shinshu) in Kyoto, Japan. His research interests cover institutional aspects of crop genetic resources management, participatory agricultural research and inter-civic relations. He worked for the Graduate School of International Development at Nagoya University as the director of Rural and Regional Development Management Program from 2008 to 2013. He also worked at Ethiopian Institute of Agricultural Research as an expert in appropriate technology development from 2012 to 2015. He is a member of the councils of the Japanese Society of International Development and the Japan Society of Organic Agriculture.

Michel Pimbert is Professor of Agroecology and Food Politics as well as Director of the Centre for Agroecology, Water and Resilience at Coventry University, UK. He was formerly a member of the High Level Panel of Experts on Food Security and Nutrition of the Committee on World Food Security at the Food and Agriculture Organization of the United Nations. His research interests include agroecology and food sovereignty; the political ecology of biodiversity and natural resource management; participatory action research methodologies; and deliberative democratic processes. He works with networks of small and family farmers, Indigenous peoples and communities to advance transdisciplinary and transformative ways of knowing that regenerate local ecologies, economies and cultural diversity.

Khadija Razavi as a specialist in community participation, biodiversity conservation, women in sustainable development and appropriate technology.

Maedeh Salimi is a member of the board of the Centre for Sustainable Development and Environment (CENESTA)s.

Maria Scholten has worked as a researcher and project manager, in the field of Scottish landraces and small-scale farming.

Mya Shew is a retired staff of Yezin Agricultural University, Myanmar.

Krystyna Swiderska is a Principal Researcher at the International Institute for Environment and Development (IIED), leading IIED's work on biocultural heritage. She has conducted research with Indigenous Peoples for over 20 years, with a particular focus on traditional knowledge, customary laws, crops, and biocultural territories.

Norie Tamura is a senior researcher at the Research Institute for Humanity and Nature, Kyoto, Japan. Her research interests cover forestry and fishery resource management, sustainable agrifood transitions, and commons and commoning studies.

Min San Tein and Ohm Mar Saw work for Ministry of Agriculture, Livestock and Irrigation, Myanmar.

Mitsuyuki Tomiyoshi is an Associate Professor at Kurume University, Japan. Research areas cover NPOs (Non-Profit Organizations), social business, agriculture & agricultural communities, and paddy & Satoyama conservation.

Adam Veitch is an active crofter in Lochaber, Scotland.

Kazuo Watanabe is a professor at University of Tsukuba, Japan.

LIST OF FIGURES

List of Tables

List of Boxes

CHAPTER 1

Introduction: Thinking About Seeds

Michel Pimbert

Abstract Seed diversity is crucial to the sustainability of food and agricultural systems. Yet as Michel Pimbert's survey of the global 'state of seeds' reveals, both wild and domesticated varieties are disappearing under an onslaught of human-driven pressures. Planetary crises—the sixth great extinction and climate change—constitute one. Industrialized agriculture is another: just three crops (maize, rice and wheat) currently supply over 60% of the calories humanity obtains from food. The impacts of this impoverishment on small and Indigenous farmers, ecosystems, food security and human health are manifold, and understanding them demands that we unravel a range of intermeshed social and political factors. Disparities in wealth, gender and ethnicity, for instance, determine the way seeds are cultivated, conserved, collected and exchanged. And the primary domains of seed governance—state, corporate and farm—wield different, often unequal powers. By confronting these complexities, Pimbert asserts, we can map ways of managing seeds equitably, to support human and planetary wellbeing.

M. Pimbert (✉)
Centre for Agroecology, Water and Resilience, Coventry University, Coventry, UK
e-mail: michel.pimbert@coventry.ac.uk

© The Author(s) 2022
Y. Nishikawa and M. Pimbert (eds.), *Seeds for Diversity and Inclusion*,
https://doi.org/10.1007/978-3-030-89405-4_1

Keywords Biodiversity loss · Indigenous peoples · Farmers · Adaptive management of diversity · Seed governance

Sustainable food and agriculture depend on the continued availability and quality of, and access to, seeds of cultivated and wild plants for renewal and adaptation to dynamic change. Different types of seed biodiversity ('cultivated', 'reared' or 'wild') are used by different people at different times and in different places, and so contribute to ecological sustainability, food security and livelihood strategies in a complex manner. Understanding how cultivation, management, collection, use and marketing of different types of domesticated and wild seeds are affected by differences in wealth, gender, race, ethnicity and age is essential for making equitable decisions on how to conserve, exchange and use seeds for human and planetary well-being.

And diversity in the seeds of cultivated and wild plants is indeed key in this context. Domesticated, semi-wild and wild seed plants are closely associated in most food- and fibre-growing environments. For example, rural and urban home gardens are typically structurally complex, and provide many multifunctional benefits to surrounding ecosystems and to people. Evidence shows that high levels of inter- and intra-specific diversity in cultivated and wild seed plants—especially locally adapted landraces and wild crop relatives—are preserved in home gardens throughout the world (Galluzzi et al., 2010).

The interactions between the environment, seeds and local management practices also influence evolutionary processes such as introgression from wild relatives, hybridization between cultivars, mutations and natural and human selections. This generates a diversity of seeds (landraces and wild ecotypes) that are well adapted to the mosaic of changing local environmental conditions and community preferences.

As one component of agricultural biodiversity (see Box 1.1), seed diversity is thus vitally important for the design of sustainable agroecosystems and just food systems[1] (FAO, 1998; Mulvany, 2020; Peschard & Randeria, 2020; Pimbert, 1999).

Box 1.1: Agricultural biodiversity

Agricultural biodiversity (or agrobiodiversity) refers to the variety and variability of animals, plants and micro-organisms key to food and agriculture, and which result from the interaction between the environment, genetic resources and human management systems and practices. The term takes into account not only genetic, species and agroecosystem diversity and the different ways in which land and water resources are used for production, but also cultural diversity, which influences human interactions at all levels. It has spatial, temporal and scale dimensions. It comprises the diversity of genetic resources (varieties and breeds, for example) and species used directly or indirectly for food and agriculture (including, in the definition offered by the Food and Agriculture Organization of the United Nations (FAO), crops, livestock, forestry and fisheries) for the production of food, fodder, fibre, fuel and pharmaceuticals. It also covers the diversity of species that support production (such as soil biota, pollinators and predators) and those in the wider environment that support agroecosystems (agricultural, pastoral, forest and aquatic), as well as the diversity of the agroecosystems themselves.

Source FAO (1998)

This book focuses on the governance and management of cultivated and wild seeds in diverse contexts. Drawing on case studies from Japan, Taiwan, Bhutan, Myanmar, Iran, Italy, Peru and Scotland, this collection of papers offers a nuanced view on practices that exist alongside a continuum of informal to formal processes—from local to global. The main message of this book is that seed governance and management need to be transformed and based on principles of decentralization, dynamic adaptation and cultural and spiritual diversity, as well as democracy and inclusion.

---

[1] A food system gathers all the elements (environment, people, inputs, processes, infrastructures and institutions, for example) and activities that relate to the production, processing, distribution, preparation and consumption of food, and the outputs of these activities, including socioeconomic and environmental outcomes (HLPE, 2014).

1.1   Diverse Seeds Under Threat

Over the last 60 years, numerous scientific reports have documented the rapid loss of biodiversity important for food and agriculture (FAO, 2019; IPBES, 2019). The Food and Agriculture Organization of the United Nations (FAO) reported:

> *Soil, water, and genetic resources constitute the foundation upon which agriculture and world food security are based. Of these, the least understood and most undervalued are plant genetic resources. They are also the resources most dependent upon our care and safeguarding. And they are perhaps the most threatened.* (FAO, 1996)

Most notably, the global expansion of genetically uniform monocultures of industrial farming has accelerated the erosion of seed diversity. According to FAO, 75% of plant genetic diversity has been lost, as farmers worldwide abandon their locally adapted crop varieties for the genetically uniform, high-yielding varieties promoted by industrial and Green Revolution-influenced agriculture (FAO, 2004).

While regularly updated, this expert knowledge on the extent of genetic erosion in seeds is nevertheless incomplete. First, much of this academic and policy literature on food and agriculture focuses mainly on seeds of a few domesticated plants out of the handful of commercially valuable commodity crops in world trade.[2] Far less research and development and policy attention has been given to the seeds of the many locally and nationally important food crops in different regions of Africa, Asia, the Americas, Polynesia and Europe (FAO, 2010; Prescott Allen & Prescott Allen, 2018). This institutional bias is even more striking in light of the fact that out of the 250,000 identified and described plant species, some 30,000 are edible and around 7000 have been cultivated or collected for food at one time or another (Harlan, 1995).

The true extent and impact of the loss of cultivated and wild plant seeds on myriad peasant and Indigenous communities are unnoted by

---

[2] Only 6 crops—wheat, soybeans, maize, rice, barley and rapeseed—cover 50% of arable land, and only 9 crops account for 66% of total crop production (FAO, 2019). About 100 species contribute 90% of all calories in the human diet (Hufford et al., 2019) and three (rice, wheat and maize) represent about 60% of calories and 56% of proteins from plants consumed globally, while using nearly 50% of irrigation water.

many development professionals, scientists and inter-governmental organizations. Outsiders' expert knowledge often fails to grasp the complex, diverse and risk-prone local realities of indigenous and peasant communities (Chambers, 1997, 2017). Simply put, the seed diversity that matters to so many Indigenous and peasant farmers as well as to small- and medium-scale producers is not a priority for national and international[3] agricultural research and development.

1.2   The Unprecedented Challenges Caused by Seed Extinctions

The loss of seed diversity on farmlands and the commons is a feature of the sixth mass extinction of biodiversity the world is currently experiencing (IPBES, 2021). This irreversible loss of inter- and intra-specific crop seed diversity has created several existential threats and major challenges, as follows.

1.2.1   Growing Malnutrition and Food Insecurity

Just 30 crops supply 95% of the calories we obtain from food, while only 4 crops—maize, rice, wheat and potatoes—supply over 60%. Although the diversity of processed foods available in supermarkets and local shops seems remarkable, it is all, in truth, based on a handful of staple crops. The food industry endlessly re-engineers and recombines these into a huge array of these products. Ingredients such as high-fructose corn syrup, palm oil, refined flour, sugar and soy appear repeatedly in the ultra-processed foods that give the illusion of dietary diversity in the global food system (HLPE, 2017).

This unprecedented and ongoing reduction in dietary richness is having a significant impact on human health worldwide. For example, the decline in seed and dietary diversity is linked to a drop in human gut microbiota, the community of micro-organisms living in the gastrointestinal tract. Many of the common pathologies of the twenty-first century—inflammatory bowel disease, type 2 diabetes and obesity, for example—are associated with a reduction in microbiotic richness

---

[3] The Consultative Group for International Agricultural Research (CGIAR) and its 15 international centres have a mandate to work on a relatively small number of commodity crops: https://www.cgiar.org/research/research-centers/.

(Heiman & Greenway, 2016). By contrast, healthy individuals with resilient auto-immune systems have highly diverse gut microbiota.

Food and farming practices based on a significant diversity of wild and cultivated seed plants can increase dietary diversity and thereby improve human health by encouraging species-rich gastrointestinal microbiomes. The devastating impacts of COVID-19 and the enduring pandemic have also underscored the importance of diverse nutritious diets to strengthen peoples' immune system (Naja & Hamadeh, 2020; Yeoh et al., 2021). More generally, equitable access to, and use of, a diversity of seeds of cultivated and wild plants is also a key condition for food and nutrition security in rural and urban contexts (Pimbert & Lemke, 2018).

1.2.2   Unsustainable Food and Agricultural Systems

The International Assessment for Knowledge and Agricultural Science for Development (IAASTD + 10, 2020) and other scientific studies (HLPE, 2019; IPCC, 2019) highlight the urgent need to shift from industrial uniformity to biodiversity-rich farming in the face of increasingly rapid climate change, market volatility and potential near-future pandemics. Industrial food and farming are more than ever unsustainable because all relevant biophysical indicators are turning negative, fast, steeply, dangerously; the emerging context is beyond human experience; and the costs of mitigation, adaptation and remediation are rising sharply.

Industrial agriculture's lack of resilience to shocks and stresses is striking. The uniform monocultures of industrial farming and that based on benefits of the Green Revolution are particularly susceptible to global warming and associated changes in rainfall patterns, hurricane frequency and incidence of pest attacks. The Intergovernmental Panel on Climate Change (IPCC) points to the need to substantially increase crop genetic diversity to enable adaptation to climate change in the coming decades (IPCC, 2019, 2022).

Indeed, adaptation to climate change and recovery from climate-induced disasters hinges on the availability and free access to a diversity of seeds needed to rediversify farming systems for socioecological resilience (Chapter 8). In this regard, agroecology offers viable avenues for mitigation of and adaptation to climate impacts (FAO, 2018; HLPE, 2019). Rooted in endogenous development visions (see Box 1.2), agroecology seeks to diversify agroecosystems by combining many different crop varieties and species into functional wholes that operate like ecosystems, reducing insect pest and diseases, recycle nutrients, conserve soils and water and adapt to climate change (Altieri, 1995; Gliessman, 2015). East Asian agroecologists are now re-discovering the endogenous farmer practices that sustained food and agriculture for centuries in countries like Japan, Korea and China (King, 1911). However, state spending for research and development (R&D) continues to massively support industrial agriculture, a high emitter of greenhouse gases.[4] Worldwide, there is a chronic lack of investment in research for biodiversity-conserving agroecology, both domestically and through overseas aid.[5]

Box 1.2: Endogenous development theory

In 1980s Japan, the sociologist Kazuko Tsurumi was among those developing a comprehensive theory of endogenous development, or 'development from within', as an alternative to ideas of modernization originating in Europe and the United States.

The basic goal of endogenous development as Tsurumi envisaged it is for all humans and communities to meet needs in food, clothing, shelter and medical care, as well as to create conditions in which individuals can fully achieve their potential. But the paths to that goal follow diverse processes of social change. Individuals and groups in each region must autonomously create social visions and ways forward to the goal by adapting to their own ecological systems, and by basing development programmes on their own cultural heritage and traditions.

Source Tsurumi, K. Aspects of Endogenous Development in Modern Japan Research Papers, Series A-36. Institute of International Relations, Sophia University, Tokyo, 1979

---

[4] The global food system is responsible for at least 37% of total net anthropogenic greenhouse-gas emissions (IPCC, 2019).

[5] In the United States, for example, a recent analysis of funding by the US Department of Agriculture (USDA) showed that projects with an emphasis on agroecology represented only 0.6–1.5% of the entire 2014 USDA Research, Extension, and Economics budget (Delonge et al., 2016). UK development aid barely supports agroecology: overseas aid for agroecological projects in Africa, Asia and Latin America accounts for less than 5% of agricultural aid and less than 0.5% of the total UK aid budget since 2010 (Pimbert & Moeller, 2018). Similarly, EU funding to FAO and other Rome-based UN agencies and the Green Climate Fund neglects agroecological R&D (Moeller, 2020); the lion's share of funding goes to industrial agriculture, which is responsible for most greenhouse gas emissions.

1.2.3   Loss of Ecosystem Functions, Goods and Services

From pollination to natural pest control and water purification, ecosystems provide key functions, goods and services, and they too depend on promoting inter- and intra-specific seed diversity on farms and surrounding landscapes (IPBES, 2019). Diverse seed-producing wild and cultivated plants are centrally involved in the mediation of multiple ecological functions and processes, at different scales (Pimbert, 1999). Valuable ecological processes that result from the interactions between species, and between species and the environment, include biogeochemical cycling, the maintenance of soil fertility and water quality, providing food and refuges for pollinators, and regulating climate. The erosion of seed diversity—and the ensuing loss or reduction in the abundance of cultivated and wild plants in space and time—fundamentally undermines vitally important ecological functions and processes that sustain and renew the material basis of social and economic life (IPBES, 2019).

Appropriate responses to these unprecedented existential threats require a radical transformation in the governance of seeds, and the systems in which they are embedded.

1.3   The Politics of Seed Governance

Decisions on how, why, where and by whom cultivated and wild plant seeds are conserved, exchanged and used are critical for the future of food and agriculture as well as the well-being of people and planet.

Seed governance is defined here as the set of political, social, economic and administrative rules, processes and systems that determine the way decisions by the various actors are taken and implemented for the management and use of seeds. Governance also includes the rules and processes through which decision-makers are held accountable locally, nationally and internationally.

Complex local livelihoods and food provisioning depend on cultivated, semi-wild and wild plants in most rural and peri-urban communities across the world (Guijit et al., 1995). Governance therefore needs to be comprehensive and inclusive in its focus on seeds of both cultivated and wild plants important for food and agriculture.

1.3.1   Domesticated/Cultivated Seed Plants

Governance centres here on the conservation of crop seeds (in situ and ex situ), seed multiplication, seed hygiene, seed certification and catalogues, plant breeding, distribution and exchange of seeds, Indigenous and peasant knowledge, informal seed exchange networks, collective and customary rights, plant breeders' rights and private property rights, commercialization of seeds and seed corporations (see Chapters 4, 5, 6, 9, 10 and 12). The most contested issues in national seed laws and policies are on peasants' and farmers' rights to save, use, sow, re-sow, exchange, share and sell farm produce, including seeds of varieties protected by plant breeders' rights.

1.3.2   Semi-Wild/Wild Seeded Plants

Governance encompasses the management of lands where wild crop relatives and wild food plants live, including the commons (such as grasslands, forests, wetlands and drylands) where wild plants live, reproduce and scatter their seeds. These are mostly human-managed ecosystems and landscapes with long history of coevolution between people and nature. Many indigenous, pastoral and peasant communities obtain diverse foods and fibres from these anthropogenic landscapes (Gómez-Pompa & Kaus, 1992; Pimbert & Borrini, 2020) and biocultural heritage territories (see Chapter 4). Policies that restrict local access to these humanized landscapes are often contested by Indigenous and peasant communities who have historically depended on them for their livelihoods and culture (FPP, 2020; Ghimire & Pimbert, 1997).

1.4   THE ACTORS AND INSTITUTIONS GOVERNING SEEDS

There are multiple actors with contrasting powers involved in making decisions on the governance of seeds. They can be described in terms of Marc Nerfin's typology of the Prince, the Merchant and the Citizen (Nerfin, 1986).

1.4.1   The Prince: The State

Different levels of government—from national to local as well as inter-governmental international institutions—are different manifestations of the state, or as Nerfin has it, 'the Prince'. Nation states, although by no means all of them, have signed up to several international treaties and declarations key to the governance of seeds (see Box 1.3). Under these treaties and commitments, states have obligations to ensure the consistency of their national laws and policies, and of international agreements and standards to which they are party regarding the right to seeds. Here is an example from Article 19 of the United Nations Declaration on the Rights of Peasants (UNDROP):

> *States shall ensure that seed policies, plant variety protection and other intellectual property laws, certification schemes and seed marketing laws respect and take into account the rights, needs and realities of peasants and other people working in rural areas.*[6]

However, the implementation of joined-up and consistent approaches at national and local levels is problematic for most governments. Siloed administrations, top-down interventions, structural constraints and sectoral approaches often hamper coordinated and relevant action on the ground. For example, the governance and management of cultivated and wild seeds are generally the responsibility of separate ministries, reflecting an enduring opposition between development and conservation. In turn, this often leads to a mismatch between standard conservation and development approaches and the multitude of diverse local realities and needs of communities dependent on both cultivated and wild diversity (Chambers, 1997; Ghimire & Pimbert, 1997; Scott, 2020).

---

[6] The legally binding International Treaty on Plant Genetic Resources for Food and Agriculture (ITPGRFA) states that contracting parties should, as appropriate, and subject to its national legislation, take measures to protect and promote farmers' rights, including: protection of traditional knowledge relevant to plant genetic resources for food and Agriculture; the right to equitably participate in sharing benefits arising from the utilization of plant genetic resources for food and agriculture; and the right to participate in making decisions, at the national level, on matters related to the conservation and sustainable use of plant genetic resources for food and agriculture.

Box 1.3: The major institutions of seed governance

The Convention on Biological Diversity (CBD). Adopted in 1992, the CBD protects important elements of peasants' right to seeds, including through provisions to ensure the protection of Indigenous and local communities' traditional knowledge, and the fair and equitable sharing of the benefits arising out of the utilization of genetic resources[7]

The International Treaty on Plant Genetic Resources for Food and Agriculture (ITPGRFA). This is the most important international treaty for the recognition and protection of farmers' and peasants' right to seeds. Its preamble states that 'the rights recognized in this Treaty to save, use, exchange and sell farm-saved seed and other propagating material, and to participate in decision-making regarding, and in fair and equitable sharing of the benefits arising from, the use of plant genetic resources for food and agriculture, are fundamental to the realization of Farmers' Rights, as well as to the promotion of Farmers' Rights at national and international levels'. The responsibility for realizing farmers' rights rests with national government, as stated in the treaty.[8]

The United Nations Declaration on the Rights of Indigenous Peoples (UNDRIP). This declaration recognizes Indigenous peoples' right to maintain, control, protect and develop their own seeds and their ownership of those seeds.[9]

The Agreement on Trade-Related Aspects of Intellectual Property Rights (TRIPS). Adopted by the World Trade Organization (WTO). Member States of the WTO must protect intellectual property rights over plant varieties either by patents, an effective *sui generis* system (a system of its own kind) or a combination of both. Patents are the most comprehensive form of protection that can be granted because they give the right-holders—in many cases corporations—exclusive rights over plant-related inventions.[10]

---

[7] https://www.cbd.int/convention/.

[8] http://www.fao.org/3/i0510e/i0510e.pdf.

[9] https://www.un.org/development/desa/indigenouspeoples/declaration-on-the-rights-of-indigenous-peoples.html.

[10] https://www.wto.org/english/docs_e/legal_e/27-trips_03_e.htm.

International Union for the Protection of New Varieties of Plants (UPOV) and its Convention (UPOV Convention). The UPOV Convention protects the rights of plant breeders who have developed varieties that are new, distinct, uniform and stable (DUS). UPOV requires peasants to obtain authorization to sell protected seeds. This practically prohibits the realization of farmers' rights. Moreover, the UPOV 1991 Act further prohibits farmers from saving, reusing, and exchanging these seeds (except in a very limited way on their own farms).[11]

The Voluntary Guidelines on the Responsible Governance of Tenure of Land, Fisheries, and Forests in the context of National Food Security (VGGT). Agreed by the UN Committee on World Food Security, it promotes responsible governance of land, forests and fisheries under all forms of tenure: public, private, communal, indigenous, customary and informal.[12]

The United Nations Declaration on the Rights of Peasants and Other People Working in Rural Areas (UNDROP). This is the most recent UN instrument that recognizes new human rights—including the right to seeds, land, natural resources and food sovereignty via agroecology, local seeds, local markets, gender equity and participatory decision-making.[13]

1.4.2   The Merchant: Seed Corporations

Over the last 60 years, the commercial seed sector has become increasingly consolidated and concentrated. Many small and family-owned seed companies have been absorbed into larger seed firms through mergers and acquisitions. The most recent mergers reduced the number of major seed companies to four: Bayer-Monsanto, DowDuPont/Corteva, ChemChina-Syngenta and BASF. This handful of corporations control more than 60% of global proprietary seed sales (Howard, 2020). Agriculture in industrialized countries sources most of its seeds from them. In other regions of the world, the proportion of commercial seeds used

---

[11] https://www.upov.int/portal/index.html.en.

[12] http://www.fao.org/tenure/voluntary-guidelines/en/.

[13] https://www.geneva-academy.ch/joomlatools-files/docmanfiles/UN%20Declaration%20on%20the%20rights%20of%20peasants.pdf.

is still relatively low—approximately 30% in India, and less than 10% in Africa.

Unlike many family seed companies and cooperatives (Chapter 10), transnational seed corporations strongly promote a controllable uniformity through proprietary technologies such as hybrid seeds that meet DUS[7] criteria, patented genetically modified (GM) seeds and gene drive technologies, and more generally through discourse, policy influence and private–public R&D partnerships that further their political and economic interests (Clapp & Fuchs, 2009).

As the so-called Fourth Industrial Revolution (4IR) for agriculture gains momentum (WEF, 2018), global seed corporations focus on rolling out 4IR technologies[8] designed to transform food and agriculture. To this end, they are building partnerships with other agro-industrial giants in a range of arenas, from artificial intelligence, robotics, digital sequencing, synthetic biology, big data, pesticides, farm machinery, e-commerce, investment finance and private equity. In the meantime, their use of pro-business investor dispute panels allows seed corporations to block government interventions to regulate their activities, as well as to seek financial compensation for lost market opportunities and investments. In effect, investor-state dispute settlement (ISDS) is a system through which investors can sue countries for alleged discriminatory practices.[9]

In this unprecedented thrust to further enclose the commons, corporate seed governance promotes ever-increasing uniformity, privatization, financialization, control, centralization and coercion (Aubry, 2019; FIAN, 2020; Hache & Spash, 2021; IPES Food & ETC, 2021).

---

[7] Under the International Union for the Protection of New Varieties of Plants (UPOV) rules, a new plant variety must comply with the requirements of distinctness, uniformity and stability (DUS) in order to be registered and protected.

[8] The Fourth Industrial Revolution (4IR) for food and agriculture is based on a package of 12 technologies, including precision agriculture to 'optimize the use of agricultural inputs and water', gene editing, big data and advanced analytics, the 'Internet of Things' for real-time traceability of the food chain, alternative proteins, and nutrigenetics for personalized nutrition (WEF, 2018).

[9] For examples and more information see http://isds.bilaterals.org/the-basics#.

1.4.3   The Citizen: Food Producers and Consumers

There are more than 570 million farms worldwide, most of which are small and family farms (Lowder et al., 2016). Of these, 74% are located in Asia, with China alone representing 35% and India 24% of all farms. Some 72% of the world's farms are smaller than 1 hectare (ha) in size; 12% are 1 to 2 ha in size (small farms); and 10% are between 2 and 5 ha. Only 6% of the world's farms are larger than 5 ha. Family farmers work 75% of the world's agricultural land and are responsible for most of the world's food and agricultural production (Lowder et al., 2016).

As historical custodians of the land, small farmers have co-created, with nature, myriad locally adapted seeds. In Southeast Asia, for instance, the high diversity of ethnic groups within a relatively small region has produced extraordinary diversity in Indigenous vegetables, as different groups favour specific culinary and agronomic properties (Gill et al., 2013). Farm-saved seed and informal seed exchanges are common practices among small and family farmers, with informal seed systems providing 60 to 100% of seeds planted by Indigenous and peasant communities in the Global South (Almekinders & Louwaars, 2002). These small-scale producers conserve, share, and use diverse seeds through their decentralized governance and adaptive practices (Brac de la Perrière, 2014; IPC, 2017; Peschard & Randeria, 2020).

The US farmer, environmental activist and poet Wendell Berry has said that 'eating is an agricultural act' (Berry, 1990). Every human thus plays a direct and indirect role in enabling (or undermining) seed diversity in agri-food systems. People's decisions to source food locally generally help to conserve and enhance seed diversity in short food chains and local food systems—for example, in Scotland (Chapter 9) and other parts of Europe (Kneafsey et al., 2013). By contrast, people's ability to support seed diversity is much more limited when they rely on long-distance value chains based on uniformity and economies of scale. This is because food choices—and which seeds are ultimately conserved and used—are largely determined by distant corporations that control the different links of these global value chains, from seeds and farm inputs to industrial food processing and supermarkets (HLPE, 2017).

Nerfin's third system, the Citizen, emphasizes autonomy and the need for citizens to self-organize and self-govern in local settings.[10]

---

[10] https://www.daghammarskjold.se/publication/another-development-third-system/.

Throughout the world, many Indigenous and peasant communities still develop their own place-specific seed governance and management rules. Mutual agreements on the roles, rights and obligations of different local actors allow them to adaptively govern and manage their seed commons (Chapters 2, 4, and 12; Borrini-Feyerabend et al., 2007).

The Citizen can also exert power from below to change actions taken by the Prince or the Merchant (Chapter 12). The collective agency and power of citizens partly depends on their capacity to educate, mobilize for autonomous action through horizontal networks and organize to change policies and institutions to reflect their own priorities and cosmovisions. This implies radical changes in power relations and people's self-determination in the governance and management of seeds, as advocated by movements for endogenous development (Chapters 3 and 13; Kato, 2020) and food sovereignty (Chapter 2).

By considering various international examples and local intitiatives, this book highlights the collective capacity of a growing international movement to reclaim seeds for diversity and autonomy in food and farming.

References

Almekinders, C. J. M., & Louwaars, N. P. (2002). The importance of the farmers' seed systems in a functional national seed sector. *Journal of New Seeds, 4*(1–2), 15–33. https://doi.org/10.1300/J153v04n01_02

Altieri, M. A. (1995). *Agroecology: The science of sustainable agriculture*. Boulder, Colorado: Westview Press.

Aubry, S. (2019). The future of digital sequence information for plant genetic resources for food and agriculture. *Frontiers in Plant Science, 10*(1046). https://doi.org/10.3389/fpls.2019.01046

Berry, W. (1990). The pleasures of eating. In W. Berry, *What are people for?* North Point Press.

Borrini-Feyerabend, G., Pimbert, M. P., Farvar, M. T., Kothari, A., & Renard, Y. (2007). *Sharing power: A global guide to collaborative management of natural resources* (2nd ed.). Earthscan, IIED and IUCN.

Brac de la Perrière, R. A. (2014). *Semences paysannes, plantes de demain* [*Peasants' seeds, plants of tomorrow*]. Charles Léopold Mayer.

Chambers, R. (1997). *Whose reality counts? Putting the last first*. Intermediate Technology Development Group.

Chambers, R. (2017). *Can we know better? Reflections for development*. Practical Action Publishing.

Clapp, J., & Fuchs, D. (2009). *Corporate power in global agrifood governance.* MIT Press.

DeLonge, M. S., Miles, A., & Carlisle, L. (2016). Investing in the transition to sustainable agriculture. *Environmental Science & Policy, 55*(1), 266–273. https://doi.org/10.1016/j.envsci.2015.09.013

FAO. (1996). *The state of the world's plant genetic resources for food and agriculture.* Rome, FAO.

FAO. (2004). Erosion of plant genetic diversity. *FAO News.* Retrieved December 28, 2021, from https://www.fao.org/Newsroom/en/focus/2004/51102/article_51107en.html

FAO. (1998). *International Technical Workshop organized jointly by the Food and Agriculture Organization of the United Nations and the Secretariat of the Convention on Biological Diversity (SCBD), with the support of the Government of the Netherlands 2–4 December 1998.* www.fao.org/sd/epdirect/EPre0063.htm

FAO. (2010). *The second report on the state of the world's plant genetic resources for food and agriculture.* FAO.

FAO. (2019). *The state of the world's biodiversity for food and agriculture,* J. Bélanger & D. Pilling (Eds.). FAO Commission on Genetic Resources for Food and Agriculture Assessments. FAO.

FIAN. (2020). *Rogue capitalism and the financialization of territories and nature.* P. Seufert, R. Herre, S. Monsalve & S. Guttal (Eds.). FIAN International, Transnational Institute and Focus on the Global South.

FPP. (2020). *Human rights in the post-2020 Global Biodiversity Framework: options for integrating a human-rights based approach to achieve the objectives of the Convention on Biological Diversity.* Forest Peoples Programme. Retrieved July 1, 2021, from https://www.forestpeoples.org/sites/default/files/documents/humanrights%20Eng_0.pdf

Galluzzi, G., Eyzaguirre, P., & Negri, V. (2010). Home gardens: Neglected hotspots of agro-biodiversity and cultural diversity. *Biodiversity Conservation, 19,* 3635–3654. https://doi.org/10.1007/s10531-010-9919-5

Ghimire, K. B., & Pimbert, M. P. (1997). *Social change and conservation: Environmental politics and impacts of national parks and protected areas.* Earthscan/Routledge.

Gliessman, S. R. (2015). *Agroecology: The ecology of sustainable food systems.* Boca Raton, Florida: CRC Press.

Gill, T. B., Bates, R., Bicksler, A., Burnette, R., Ricciardi, V., & Yoder, L. (2013). Strengthening informal seed systems to enhance food security in Southeast Asia. *Journal of Agriculture, Food Systems, and Community Development, 3*(3), 139–153. https://doi.org/10.5304/jafscd.2013.033.005

Gómez-Pompa, A., & Kaus, A. (1992). Taming the wilderness myth: Environmental policy and education are currently based on Western beliefs about nature rather than on reality. *BioScience, 42*(4), 271–279. https://doi.org/10.2307/1311675

Guijit, I., Hinchcliffe, F., Melnek, M., Bishop, J., Eaton, D., Pimbert, M. P., Pretty, J., & Scoones, I. (1995). *The hidden harvest: The value of wild resources in agricultural systems.* IIED.

Hache, F., & Spash, C. L. (2021). *Nature, life & relations–'Optimised': A policy brief on the Dasgupta Review.* Green Finance Observatory. Retrieved July 19, 2021, from https://greenfinanceobservatory.org/2021/05/26/the-dasgupta-review-deconstructed/

Harlan, J. R. (1995). *The living fields: Our agricultural heritage.* Cambridge University Press.

Heiman, M. L., & Greenway, F. L. (2016). A healthy gastrointestinal microbiome is dependent on dietary diversity. *Molecular Metabolism, 5,* 317–320. https://doi.org/10.1016/j.molmet.2016.02.005

HLPE. (2014). *Food losses and waste in the context of sustainable food systems.* A report by the High Level Panel of Experts on Food Security and Nutrition of the Committee on World Food Security, Rome.

HLPE. (2019). *Agroecological and other innovative approaches for sustainable agriculture and food systems that enhance food security and nutrition.* A report by the High Level Panel of Experts on Food Security and Nutrition of the Committee on World Food Security, Rome.

HLPE. (2017). *Nutrition and food systems: A report by the High Level Panel of Experts on Food Security and Nutrition of the Committee on World Food Security.* CFS.

Howard, P. H. (2020). How corporations control our seeds. In S. Jarayaman & K. De Master (Eds.), *Bite back: People taking on corporate food and winning.* University of California Press.

Hufford, M. B., Mier, B., Teran, J. C., & Gepts, P. (2019). Crop biodiversity: An unfinished magnum opus of nature. *Annual Review of Plant Biology, 70,* 727–751. https://doi.org/10.1146/annurev-arplant-042817-040240

IAASTD +10. (2020). Transforming our food systems. In H. R. Herren & B. Haerlin (Eds.), *The IAASTD+10 advisory group.* Zurich: Biovision.

IPBES. (2019). *Summary for policymakers of the global assessment report on biodiversity and ecosystem services of the Intergovernmental Science-Policy Platform on Biodiversity and Ecosystem Services.* IPBES.

IPBES. (2021). *Nature's dangerous decline 'unprecedented'; Species extinction rates 'accelerating'.* Media Release. IPBES. Retrieved July 1, 2021, from https://ipbes.net/news/Media-Release-Global-Assessment

IPC. (2017). *Peasants give life to biodiversity.* IPC. Retrieved July 3, 2021, from http://www.foodsovereignty.org/wp-content/uploads/2016/10/ENGLISH_spreads_lowRes_.pdf

IPCC. (2019). *Climate change and land.* Special Report: Summary for Policymakers. IPCC. Retrieved July 19, 2021, from https://www.ipcc.ch/srccl/

IPCC. (2022). *Climate change 2022: Impacts, adaptation, and vulnerability.* New York: Working Group II contribution to the Sixth Assessment Report of the intergovernmental panel on climate change.

IPES Food & ETC. (2021). *A long food movement: transforming food systems by 2045.* IPES Food.

Kato, M. (2020). Minakata Kumagusu: The first Japanese environmentalist. *Educational Philosophy and Theory.* https://doi.org/10.1080/00131857.2020.1770612

King, F. H. (1911). *Farmers of forty centuries: Permanent agriculture in China, Korea, and Japan.* Rodale Press.

Kneafsey, M., Venn, L., Schmutz, U., Balázs, B., Trenchard, L., Eyden-Wood, T., et al. (2013). *Short food supply chains and local food systems in the EU: A state of play of their socio-economic characteristics.* Publications Office of the European Union.

Lowder, S. K., Skoet, J., & Raney, T. (2016). The number, size, and distribution of farms, smallholder farms, and family farms worldwide. *World Development, 87,* 16–29.

Moeller, N. I. (2020). *Analysis of funding flows to agroecology: The case of European Union monetary flows to the United Nations' Rome-based agencies and the case of the Green Climate Fund.* CIDSE & CAWR.

Mulvany, P. (2020). Sustaining agricultural biodiversity and heterogeneous seeds. In A. Kassam & L. Kassam (Eds.), *Rethinking food and agriculture.* Woodhead Publishing.

Naja, F., & Hamadeh, R. (2020). Nutrition amid the COVID-19 pandemic: A multi-level framework for action. *European Journal of Clinical Nutrition, 74,* 1117–1121.

Nerfin, M. (1986). Neither prince nor merchant: Citizen, an introduction to the third system. In S. J. Patel (Ed.), *World economy in transition* (pp. 47–59). Pergamon Press.

Peschard, K., & Randeria, S. (2020). 'Keeping seeds in our hands': The rise of seed activism. *The Journal of Peasant Studies, 47*(4), 613–647. https://doi.org/10.1080/03066150.2020.1753705

Prescott Allan, R., & Prescott Allen, C. (2018). *Genes from the wild: Using wild genetic resources for food and raw materials.* London: Routledge (originally published in 1988).

Pimbert, M. P. (1999). *Sustaining the multiple functions of agricultural biodiversity.* IIED Gatekeeper Series, 88. IIED.

Pimbert, M. P., & Lemke, S. (2018). Using agroecology to enhance dietary diversity. *UNSCN News, 43,* 33–42.

Pimbert, M. P., & Moeller, N. I. (2018). Absent agroecology aid: On UK agricultural development assistance since 2010. *Sustainability, 10,* 505. https://doi.org/10.3390/su10020505

Scott, J. C. (2020). *Seeing like a state: How certain schemes to improve the human condition have failed.* Yale Univerity Press.

WEF. (2018). *Innovation with a purpose: The role of technology innovation in accelerating food systems transformation.* Geneva: World Economic Forum.

Yeoh, Y. K., Zuo, T., Lui, G. C.-Y., et al. (2021). Gut microbiota composition reflects disease severity and dysfunctional immune responses in patients with COVID-19. *Gut, 70,* 698–706.

CHAPTER 2

Reclaiming Diverse Seed Commons Through Food Sovereignty, Agroecology and Economies of Care

Michel Pimbert

Abstract Seed commons—the collective management of seeds and associated knowledge—is a major aim of food sovereignty, that crucial alternative to the dead end of industrialized agriculture. To reclaim the commons, explains Michel Pimbert in this wide-ranging policy analysis, we need to enable community control over growing, trading and consuming food. That will demand mutually supportive transformations in agriculture, economies, rights and political systems towards agroecology, an economics of solidarity, collective notions of property and direct democracy. Drawing on sources such as the Nyéléni Declaration on food sovereignty and the UN Declaration on the Rights of Peasants and Other People Working in Rural Areas, Pimbert outlines a radical approach to seed governance outside the capitalist and patriarchal paradigm. The proposals, while scarcely featuring in global and national fora on seed

M. Pimbert (✉)
Centre for Agroecology, Water and Resilience, Coventry University, Coventry, UK
e-mail: michel.pimbert@coventry.ac.uk

Y. Nishikawa and M. Pimbert (eds.), *Seeds for Diversity and Inclusion*,
https://doi.org/10.1007/978-3-030-89405-4_2

governance, offer a fresh framework for needed change at a time of social exclusion, poverty and deepening environmental crises.

Keywords Community control · Food sovereignty · Agroecological transformation · Diverse seed commons

2.1   INTRODUCTION

Food sovereignty—community control over how food is consumed, traded and produced—offers a normative framework for radically rethinking how seeds are governed and managed. That in turn reveals a way of exiting the dead-end of unsustainable industrial agriculture (IAASTD, 2009; Steffen et al., 2015).

Reclaiming locally controlled and diverse seed commons is an important goal for food sovereignty. Regenerating decentralized forms of governance and management of diverse seed commons can be achieved by emphasizing several dimensions of the food sovereignty paradigm: the agroecological transformation of agri-food systems, the reinvention of an economics of care and conviviality, collective tenure and gender-equitable rights to seeds and the wider systems they are embedded in, and a deepening of democracy for social and environmental justice.

These mutually supportive transformations seek to put seeds and the food systems they are part of outside capitalism and patriarchy. This is the main argument presented in this chapter.

2.2   FOOD SOVEREIGNTY AND SEEDS

Food sovereignty aims to recreate the realm of democracy and freedom by fostering the regeneration of diverse autonomous food systems in both rural and urban areas (Pimbert, 2008). It is thus grounded in the idea that farmers and other citizens—men and women—can and should govern themselves by engaging in the practice of democracy. The Declaration of the 2007 Nyéléni Forum on Food Sovereignty affirms the centrality and primacy of "peoples" in framing policies and practices for food, agriculture, environment and human wellbeing:

Food sovereignty is the right of peoples to healthy and culturally appropriate food produced through ecologically sound and sustainable methods, and their right to define their own food and agriculture systems. It puts those who produce, distribute and consume food at the heart of food systems and policies rather than the demands of markets and corporations. It defends the interests and inclusion of the next generation. It offers a strategy to resist and dismantle the current corporate trade and food regime, and directions for food, farming, pastoral and fisheries systems determined by local producers. Food sovereignty prioritizes local and national economies and markets and empowers peasant and family farmer-driven agriculture, artisanal fishing, pastoralist-led grazing, and food production, distribution and consumption based on environmental, social and economic sustainability. Food sovereignty promotes transparent trade that guarantees just incomes to all peoples as well as the rights of consumers to control their food and nutrition. It ensures that the rights to use and manage lands, territories, waters, seeds, livestock and biodiversity are in the hands of those of us who produce food. Food sovereignty implies new social relations free of oppression and inequality between men and women, peoples, racial groups, social and economic classes and generations. (Nyéléni, 2007)

Over the past two decades, food sovereignty has been discussed and defended under the leadership of La Vía Campesina[1] (Desmarais & Nicholson, 2013; Pimbert, 2019). Other social movements have also contributed to shaping the agenda around this issue. Most notably, Indigenous peoples have expanded the food sovereignty paradigm to include sacred and spiritual dimensions of life. For example, members of the Indigenous Circle during Food Secure Canada's People's Food Policy process[2] broadened the food sovereignty framework by emphasizing: "Food is sacred — food is a gift of life, not to be squandered. It cannot be commodified". While keeping within the conceptual framing

---

[1] The term food sovereignty was first brought to international attention at the World Food Summit organized by the Food and Agriculture Organization of the United Nations in 1996. It was put forward by La Vía Campesina, an international movement that coordinates organizations of small- and medium-sized producers, agricultural workers, rural women and Indigenous communities from Asia, the Americas and Europe. During the 1996 World Food Summit, La Vía Campesina presented a set of mutually supportive principles as an alternative to the world trade policies and to realize the human right to food. In their statement, *Food sovereignty: a future without hunger* (1996), they declared: "Food sovereignty is a precondition to genuine food security".

[2] See https://foodsecurecanada.org.

developed by La Vía Campesina, Indigenous peoples also tend to empha-
size food sovereignty as a right for Indigenous peoples to choose, to
cultivate and to preserve their food practices and endogenous biocultural
values (FAO, 2021).

Caring for the diversity of cultivated and wild plant seeds lies at the
heart of food sovereignty and autonomous food systems (Pimbert, 2008).
In this sense, seed sovereignty is about Indigenous peoples, peasant
farmers, seed keepers, forest dwellers and other food producers having
the capacity and right to save, grow, sell and share their seeds. It refers
to the fundamental right of people "to breed and exchange diverse open
source seeds which can be saved and which are not patented, genetically
modified, owned or controlled by emerging seed giants".[3]

2.3   Reinventing Modernity
for Diverse Seed Commons

For food sovereignty advocates, ideas about seeds need to be liberated
from today's dominant vision of modernity and the corporate enclosure
of the commons. Reinventing modernity is necessary as a way of exiting
capitalism, and to enable a diversity of place-specific seed commons for
autonomy and endogenous development.

Throughout the world, peoples—especially youth—are affirming other
visions on how to live with, and care for, diverse seeds and the land.
Their pluralistic visions of modernity increasingly reject the commodification of nature and social relations (Rist, 2013) and focus on the
creation and maintenance of "the good life"—concepts and practices
such as *buen vivir* or *sumak kausai* in Latin America, ecological *swaraj*
in India (Kothari et al., 2014) de-growth in Europe (D'Alisa et al.,
2014; Latouche, 2011) and feminist subsistence perspectives (Mies &
Bennholdt Thomsen, 1999). In this reimagined pluriverse (Kothari et al.,
2019), ideas, discourses and practices reconnect individuals with nature
and help rebuild strong communities embedded in specific ecosystems
and their diverse seed commons.

In practice, regenerating seed commons partly depends on respect-
fully relating to seeds as sisters, mothers and living sentient beings rather

---

[3] See PBS feature The Lexicon of Sustainability https://www.pbs.org/food/features/
lexicon-of-sustainability-seeds/.

than anonymous, inert commodities. It is noteworthy that despite the dominant narrative, which treats seeds as a mere input for farming, many Indigenous, pastoral and peasant societies continue to nurture seeds as members of their own family or as part of the larger sacred world of the Pachamama (Mother Earth) in South America (Chapter 4), or Buthali in South Asia (Community Media Trust et al., 2008). Seeds are seen as not only having a soul and identity in many of these societies; they also embody the indivisibility of nature and culture. During seed festivals such as the Watunakuy, Andean indigenous communities bless and celebrate their seeds through their songs and mantras.[4] They know from experience that if seeds are not honoured, loved and deeply cared for, then crops planted from them will be disease-prone and will not yield good harvests. Peoples' expressions of love and care for local seeds thus mediate subtle agroecologies—a process Western science is only beginning to understand (Wright, 2021).

Regenerating seed commons also depends on protection from private enclosure; collective, polycentric management; sharing of formal and practical knowledge; and collective responsibility (Sievers-Glotzbach et al., 2020, 2021). "Seed commons" are commoning-based arrangements centred on seeds, in which a community conducts de facto handling, growing, breeding and sharing of seeds (Sievers-Glotzbach et al., 2020; and see Chapter 12). In effect, the seed commons is made by commoning (Boiler & Helfrich, 2019; Ruivenkamp & Hilton, 2017).

Within the framework of food sovereignty, seed commoning is part of the day-to-day activities mediated by local organizations that serve different purposes within communities (Pimbert, 2008, 2018a), such as:

- sustaining the ecological basis of agri-food systems—including producing knowledge and joint actions for the local adaptive management of land, seeds and water, as well as the development of reliable bio-physical indicators to track and respond to change, including climate change;
- coordinating human skills, knowledge and labour to generate both use values and exchange values in the economy of the agri-food system, as well as organize economic exchanges within and between territories;

---

[4] Mujumama, or Mother Seed. See https://vimeo.com/565544165.

- governing agri-food systems—including polycentric and place-specific decisions about people's access to food and natural resources (such as land and seeds) as well as collectively generating the political knowledge needed to shape policies and institutions.

Several local organizations with different functions, powers and responsibilities are usually needed to coordinate different seed commoning activities (see Chapter 12). Such "nested organizations" operate at different scales and act in complementary ways. These interlinked organizations and networks provide the institutional landscape that is needed to manage dynamic complexity in the social and ecological realms in which seeds and food systems are embedded.

This polycentric web of interacting organizations provides the basis for decentralized governance, and autonomous seed and food systems (Pimbert, 2008). It also helps keep seeds in farmers' hands and maintains the high diversity of cultivated and wild seed plants needed to build agroecology-based agri-food systems that are resilient to climate change, pandemics and market volatility.

2.4   How Agroecology Sustains Seed Diversity

Rooted in local Indigenous and peasant knowledge and the science of ecology and complexity, agroecological practices are reliant on high seed diversity: multi-species polycultures, intercrops, agroforestry, genetic mixtures, mixed farming and agro-sylvo-pastoral systems (Altieri, 1995; Gliessman, 2015). Agroecology also works to diversify the ecosystems and landscapes in which farming systems are embedded (Pimbert et al., 2021).

For example, in the "forest home gardens" that cover 15% of the land in Sri Lanka, family farmers raise trees, shrubs, herbs, crops and animals in a complex multi-layered agroecological system. The garden system mimics and merges with the complex structure and multiple functions of a forest, although it is not identical to it. A diversity of cultivated and edible wild seed plants are combined at multiple scales to yield many benefits, including resilience to climatic shocks and stresses as well as healthy nutrition in a diverse array of fruits, vegetables, spices and medicines, fodder and staple food items (Pushpakumara et al., 2012).

Worldwide, people and nature have co-created complex, multi-layered agroecologies based on cultivated and non-cultivated seeds. While research and policy mostly focus on crop seeds, wild and semi-wild seed plants continue to be key for "society-nature co-evolution" (Norgaard & Sikor, 1995), ecological sustainability and food and livelihood security (Gujit et al., 1995). For example, agricultural and forager communities in 22 Asian and African countries (as shown by 36 studies) use an average of 90–100 species. In Ethiopia, India and Kenya, aggregate country estimates can reach 300–800 species (Bharucha & Pretty, 2010; Gujit et al., 1995).

In India, women Dalit farmers in the Medak district of Telangana eat more than 40 species of highly nutritious wild greens in different seasons. The diets of these dryland farmers include 329 species or varieties of cereals, millets, oil seeds, pulses, fruit, vegetables, wild greens, roots and tubers. Seeds, roots, leaves, flowers, fruits, gums and bark are also consumed seasonally. Knowledgeable non-literate women farmers nurture the seeds of these highly nutritious wild foods in environments they have co-created with nature: collectively managed watersheds, common lands, tree plantations and woodlands, field edges and organically manured farm plots (Salomeyesudas & Satheesh, 2009). As "spiritual caretakers and co-creators of the Maya forest" (Ford & Nigh, 2015) in Central America, Maya farmers nurture diverse seed commons through their *milpa* system, a perennial multi-cropping and multi-stage cyclical agriculture and agro-forestry system based on maize and at least 90 other Mesoamerican plants.

Such Indigenous and peasant land-use practices create mosaics of agricultural areas and patches of wild biodiversity at multiple scales (Perfecto & Vandermeer, 2017). This "natural matrix" model sustains a variety of habitats and micro-environments as well as a diversity of cultivated and wild species (such as flowering seed plants, insects, birds and mammals), many of which are edible and often key for the provision of ecosystem functions such as pollination. These territories conserve a huge diversity of cultivated and wild seed plants (see Chapter 4), and are de facto governed by Indigenous and local communities who derive livelihoods from them (Pimbert & Borrini-Feyerabend, 2019). When guided by a feminist ethics of care, new agroecological ways of organizing can emerge within these territories of life, as suggested by practices in southern Mexico (Lilia et al., 2020) and decolonial feminist movements (Milgroom, 2021).

Re-localizing an agri-food system within a specific territory can significantly enhance possibilities for using a greater diversity of cultivated and wild seed plants that are adapted to the many heterogenous environments created through agroecological practices—from micro-environments to larger landscapes, as well as new economic niches along food chains. This is one of the reasons why food sovereignty approaches lead to the development of agri-food networks that re-localize agroecological production, processing, distribution, consumption and waste recycling within territories.

Such agroecological diversification and re-localization of agri-food systems within territories demands unrestricted access to high levels of inter- and intra-specific seed diversity. High levels of genetic heterogeneity within and between species enable adaptation to a rich mosaic of place-specific social and environmental conditions. Through this, the uniformity of industrial monocultures can be reversed, and replaced by increasing diversity, micro-geographical differentiation, dynamic local adaptation and a self-organizing ecological complexity. Such agroecological regeneration of seed diversity can be observed in the evolutionary plant breeding of cereals and other crops in Iran and Italy (see Chapter 8), where crop populations with high genetic diversity are grown in ways that encourage adaptation to the environment.

Similarly, seed diversity is often enhanced within territories through agroecological farming and decentralized food webs that closely link farmers with artisan producers and local markets for new products—such as the flour, bread and beer made from Hebridean rye in Scotland (Chapter 9; see also CSM, 2016).

The shift from industrial uniformity to living diversity is further enabled by a transformative agroecology that restructures and re-territorializes food and fibre production, distribution and consumption within decentralized circular systems that mimic natural ecosystems at different scales—from individual farm plots to entire cities. This re-territorializing of agri-food systems echoes the proposals of the Russian anarchist geographer Peter Kropotkin (1898) for an agrarian-industrial mutualism, in which most economic activities are re-localized in villages mixing agricultural and industrial elements, where production is controlled by those directly engaged in it. Kropotkin's ideas on how to overcome the spatial inefficiencies of capitalist production and generate synergies between small-scale industry and agriculture are particularly relevant today. For example, they might be applied to the design of shorter supply chains that are less vulnerable than global value chains to the massive disruptions caused by pandemics (UNEP, 2020).

The building blocks for circular systems based on an agrarian-industrial mutualism do exist, and include enhancing functional biodiversity, ecological clustering of industries, recycling, and localized production and consumption in specific rural and urban territories (Jones et al., 2012; Isenhour & Reno, 2019; Pimbert, 2012). Circular systems that combine food and energy production with water and waste management not only increase the use of seed diversity over time and space. They can also reduce greenhouse-gas emissions as well as ecological and material footprints, while maintaining a good quality of life through controlled processes of de-growth in consumption and production.[5]

More generally, agroecological pathways to sustainability can help reclaim the seed commons by using a wide diversity of heterogeneous cultivated and wild seeds in agri-food systems and the environments they are embedded in. Decentralized and re-territorialized agroecological systems using a large diversity of heterogeneous seeds (cultivated and wild plants) are usually more resilient to shocks and stresses, including climate change and market volatility.[6]

However, reclaiming a diversity of seed commons through agroecology demands a system-wide change in which seed management and governance are part of a larger paradigm shift towards food sovereignty. Such large-scale agroecological transformations depend on more inclusive democracy and justice in six key domains: access to natural ecosystems, including land, water and seeds; systems of economic exchange and markets; knowledge and culture; social networks and local organizations; discourses; and equity, gender and diversity (Anderson et al., 2021).

Within each of these domains exist structures and processes that constrain agroecology, and others that enable it. Different means are deployed by specific actors, such as agri-business and civil society organizations, to ensure that agroecology either "fits and conforms" or "stretches and transforms" the dominant agri-food regime (Levidow et al., 2014). This is a highly charged political process that

---

[5] Such as those encapsulated in the eight "Rs": re-evaluate, re-conceptualize, re-structure, redistribute, re-localize, reduce, reuse and recycle (Latouche, 2011).

[6] The resilience of such biodiversity-rich agri-food systems emerges from internal processes of functional diversity and redundancy, self-regulation, connectivity, response diversity, space and time heterogeneity, the building of natural assets such as soil fertility, social self-organization, reflective learning, autonomy and local interdependency (Tittonell, 2020).

creates major controversies and power conflicts at local, national and global levels (Anderson et al., 2021; Pimbert, 2018b).

2.5   Reinventing an Economics of Care

From a food sovereignty perspective, a central challenge for seed governance and management is to claim, recover and implement economic processes that support community control over seeds, knowledge and the means of livelihoods. Two interrelated issues are key in this context.

2.5.1   Access to Land, Seeds, Water and Other Means of Production

Colonial powers, agri-business corporations, conservation organizations and national governments: all have a history of appropriating seeds, land and natural resources, and denying the pre-existing rights of Indigenous peoples and peasant communities. Mutualities of care and community solidarity are eroded as the subsistence economy of the commons is transformed into marketable "goods and services" by private enterprises that organize wage labour to meet "consumer demand" (Illich, 2005).

In response to these enclosures, food and seed sovereignty activists have defined, demanded and defended access to land, seeds, water and other means of production as a human right, and important international instruments and agreements have been achieved in the last three decades (Claeys, 2015; Golay & Bessa, 2019; Pimbert & Borrini-Feyerabend, 2019). The United Nations Declaration on the Rights of Peasants and Other People Working in Rural Areas (UNDROP)[7] is the most recent UN instrument that recognizes new human rights. These include the right to land, seeds, natural resources and food sovereignty via agroecology, local markets, local seeds, participatory decision-making, gender justice and the transition to resilient and sustainable food systems (La Vía Campesina, 2020).

Ensuring that governments enforce and protect the collective and individual rights enshrined in UNDROP, along with other international instruments and declarations, depends on the agency and collective action of peoples and communities. In the interests of equity and non-discrimination, food and seed sovereignty movements must focus on securing collective rights and promote at all levels the equitable resolution of power dynamics related to gender, wealth, age, disability, ethnic

---

[7] The UNDROP was approved by the UN General Assembly in December 2018. It is available at http://ap.ohchr.org/documents/dpage_e.aspx?si=A/HRC/39/L.16.

background and other axes of difference (Claeys and Bourke Martignoni, 2021).

2.5.2 Diverse Economies of Care for Diverse Seeds

From a food sovereignty perspective, many custodians of diverse seed commons need their own distinct forms of economic exchange that minimize the need to participate in global commodity markets. In essence, they need diverse economies to sustain their unique seed commons and autonomous food systems.

Fortunately, "more-than-capitalist economies" (Gibson-Graham & Dombroski, 2020) persist across the world. In fact, much of the world's economy is informal, cooperative, hidden, community-based and unwaged (Rist, 2011; White & Williams, 2014). Empirical examples from economic geography show how diverse economies can also include more than human labour and human/non-human interdependence (Gibson-Graham & Dombroski, 2020). Although they are ignored, devalued and undermined by mainstream economic theory, these forms of economic organization offer relevant models for food and seed sovereignty.

For example, local control over seed saving and seed sharing is usually stronger in economies that combine market activities with non-monetary forms of exchange based on barter, reciprocity, gift relations, care and solidarity (Chapter 4; Argumedo & Pimbert, 2010). Such complementary forms of local economic exchange offer alternatives to markets solely focused on money. But to advance seed sovereignty and enable a diversity of seed commons, such plural forms of economic exchange must be acknowledged, developed and strengthened.

More fundamentally, if diverse seed commons are to be defended and food and seed sovereignty supported, a radical rethink of economics is needed. Some further ideas that could enable a post-capitalist and post-patriarchal economics include:

- a guaranteed and unconditional minimum income for all men and women
- a significant drop in time spent in wage—work and a fairer sharing of jobs and free time between men and women
- wealth redistribution measures—taxing the hyper-rich and corporations as well as financial speculations to free up resources for poorer

social groups and regions, and also regenerate local ecologies and economies

- the use of alternative local currencies to retain wealth in re-territorialized economies
- a general and progressive shift to an economics of social inclusion, freedom and solidarity—based on the principle of "from each according to his/her means, to each according to his/her needs"
- economic indicators that reflect and reinforce new definitions of wellbeing such as conviviality, mutual care and frugal abundance.

Practices for diverse economies and autonomy seek to combine these processes in mutually reinforcing ways as part of what Gibson-Graham call the "generative commons" (in de Peuter & Dyer Witheford, 2010: 46).

2.6   Deepening Democracy

Food and seed sovereignty movements seek to reverse the democratic deficit and exclusion that favour the interests of powerful corporations, investors, big farmers and technocratic research institutes. But to complement, or replace, the models of representative democracy that prevail in policymaking, "direct democracy" is often needed—that is, the direct participation of citizens in democratic decision-making. This approach is democracy in its original sense, as self-governance: people deciding their individual and collective futures.

A transition to direct democracy poses major challenges. First, deepening democracy assumes that every person is competent and reasonable enough to participate in democratic politics. It also demands a shift in mindset and behaviour from that of passive taxpayers and voters. Second, active citizenship and participation in decision-making are rights that have to be claimed mainly through the agency and actions of people themselves; they are rarely granted by the state or the market.

Third, empowering Indigenous peoples, peasant farmers and other citizens in the governance of seeds and food systems, and stewardship of the ecosystems they are embedded in (such as grasslands, forests and wetlands), demands social innovations that create inclusive and safe spaces for peoples' deliberation and action; build local organizations, horizontal networks and federations to enhance peoples' capacity for voice and agency; strengthen civil society as well as gender and intersectional equity; and expand information democracy and citizen-controlled media (community radio and video film-making). Other such needed innovations would promote self-management structures at the workplace and democracy in households; encourage learning from the history of direct democracy; and nurture active citizenship (Pimbert, 2008).

Fourth among the challenges to a shift to direct democracy is that only with some material security and free time can men and women be "empowered" to think about the policies and institutions they want and how they can develop them. Free time is needed for people to fully engage in, and regularly practise, the art of participatory direct democracy. That demands radical reforms in economic arrangements like those listed in Sect. 2.5.2. Not least, deepening democracy in the governance of seeds and agri-food systems also implies greater gender justice outside of patriarchy:

> *If we do not eradicate violence towards women within the movement, we will not advance in our struggles, and if we do not create new gender relations, we will not be able to build a new society.* (La Vía Campesina, 2008)

Forms of people-centred food systems and seed autonomy—seed commons, fruit tree gardens, diverse agroecologies, re-territorialized food systems that re-embed economics in society (cf. Polanyi, 1957)—demand inclusive participation. They also require collective action to coordinate local adaptive management and governance across a wide range of food systems and associated landscapes (farmlands, forests, grasslands and beyond). So to put people at the centre of food systems and to foster seed autonomy, it is key to decentralize and re-distribute power in polycentric and horizontal webs, both in and between territories (Pimbert & Borrini-Feyerabend, 2019).

One option is democratic confederalism. This system involves a network of bodies or councils made up of citizens, with members or delegates chosen by sortition (selection as a random sample) or elected from face-to-face democratic assemblies in villages, towns and neighbourhoods of large cities (Bookchin, 2015; Öcalan, 2011). The larger and more numerous the linked federations and confederations become, the greater is their potential to democratize and decentralize the governance of food systems and their diverse agroecologies (Pimbert, 2021).

Federating and building alliances between spaces of self-governance and bottom-up decision-making has key potential for the democratic governance of seeds and the agri-food systems they are embedded in. However, urgent issues such as the climate crisis also demands engagement with national governments. That suggests a two-pronged approach:

- acting to transform the organizational structures, professional culture and practices of state governance, and a focus on enabling national and municipal governments to support bottom-up, decentralized, multi-ethnic and participatory decision-making. Such transformations demand decisive public intervention by states to limit the disproportionate power of a handful of corporations in the governance of seeds (Clapp, 2021) and the global food system (Canfield et al., 2021).
- strengthening community self-governance and management, developing grassroots horizontal networks and insisting on participatory planning, deliberative and gender inclusive processes for policy making, participatory budgeting, power-equalizing action-research and the co-creation of new knowledge. Expanding community autonomy in governing and managing the commons also depends on enabling mutual aid, collective action, and cooperation through critical popular education (Pimbert, 2018a).

2.7   Conclusion

The proposals made here are largely absent from global and national discussions on the governance and management of seeds. In fact, many policy and scientific "experts" who are locked into "business as usual" thinking about the Fourth Industrial Revolution for food and farming (WEF, 2018; UNFSS, 2021; and see Introduction, Section 1.4) would say that these are utopian ideas that pose risks to economic progress and private property. However, given the unprecedented existential threats humanity now faces beyond the climate crisis—such as serious biodiversity loss, deepening poverty and massive social exclusion—radical ideas *outside capitalism and patriarchy* are needed to reimagine and transform seed and food systems for social and environmental justice.

In this regard, the food sovereignty paradigm—with its emphasis on a transformative agroecology, the commons, direct democracy and an economics of care and solidarity—offers hope, and a framework for action.

References

Altieri, M. A. (1995). *Agroecology: The science of sustainable agriculture*. Westview Press.

Anderson, C. R., Bruil, J., Chappell, M. J., Kiss, C., & Pimbert, M. P. (2021). *Agroecology now! Transformations towards more just and sustainable food systems*. Palgrave Macmillan. https://link.springer.com/book/10.1007/978-3-030-61315-0

Argumedo, A., & Pimbert, M. P. (2010). Bypassing globalization: Barter markets as a new Indigenous economy in Peru. *Development, 53*(3), 343–349. https://doi.org/10.1057/dev.2010.43

Bharucha, Z., & Pretty, J. (2010). The roles and values of wild foods in agricultural systems. *Philosophical Transactions of the Royal Society B, Biological Sciences, 365*(1554), 2913–26. http://rstb.royalsocietypublishing.org/content/royptb/365/1554/2913.full.pdf

Boiler, D., & Helfrich, S. (2019). *Free, fair, and alive: The insurgent power of the commons*. New Society Publishers.

Bookchin, M. (2015). *The next revolution: Popular assemblies and the promise of direct democracy*. Verso.

Canfield, M., Anderson, M. D., & McMichael, P. (2021). UN Food Systems Summit 2021: Dismantling democracy and resetting corporate control of food systems. *Frontiers in Sustainable Food Systems*. https://doi.org/10.3389/fsufs.2021.661552

Claeys, P. (2015). *Human rights and the food sovereignty movement: Reclaiming control*. Routledge.

Claeys, P., & Bourke Martignoni, J. (2021). *Women are Peasants Too: Gender equality and the UN declaration on the rights of peasants*. Policy brief. Coventry: CAWR. https://www.coventry.ac.uk/globalassets/media/global/08-new-research-section/cawr/cawr-policy-briefs/women-are-peasants-too-policy-brief---14-12-21.pdf

Clapp, J. (2021). The problem with growing corporate concentration and power in the global food system. *Nature Food*. https://doi.org/10.1038/s43016-021-00297-7

Community Media Trust, Satheesh, P. V., & Pimbert, M. P. (2008). *Affirming life and diversity: Rural images and voices on food sovereignty in South India*. IIED.

CSM. (2016). *Connecting smallholders to markets: An analytical guide*. Civil Society and Indigenous Peoples' Mechanism. Retrieved July 20, 2021, from https://www.csm4cfs.org/connecting-smallholders-markets-analytical-guide/

D'Alisa, G., Demaria, F., & Kallis, G. (2014). *Degrowth: A vocabulary for a new era*. Routledge.

de Peuter, G., & Dyer-Witheford, N. (2010). Commons and cooperatives. *Affinities: A Journal of Radical Theory, Culture and Action, 4*(1): 30–56.

Desmarais, A. A., & Nicholson, P. (2013). La Vía Campesina: An historical and political analysis. In *La Vía Campesina's open book: Celebrating 20 years of struggle and hope*. Retrieved July 19, 2021, from https://viacampesina.org/en/la-via-campesina-s-open-book-celebrating-20-years-of-struggle-and-hope/

FAO. (2021). *The White/Wiphala Paper on indigenous peoples' food systems*. FAO. https://doi.org/10.4060/cb4932en

Ford, A., & Nigh, R. (2015). *The Maya forest garden: Eight millennia of sustainable cultivation of the tropical woodlands*. Routledge.

Gibson-Graham, J. K., & Dombroski, K. (2020). *The handbook of diverse economies*. Edward Elgar Publishing.

Gliessman, S. R. (2015). *Agroecology: The ecology of sustainable food systems*. CRC Press.

Golay, C., & Bessa, A. (2019). *The right to seeds in Europe: The United Nations declaration on the rights of peasants and other people working in rural areas and the protection of the right to seeds in Europe* (Briefing No. 15). Geneva Academy of International Humanitarian Law and Human Rights.

Gujit, I., Hinchcliffe, F., Melnek, M., Bishop, J., Eaton, D., Pimbert, M. P., Pretty, J., & Scoones, I. (1995). *The hidden harvest: The value of wild resources in agricultural systems*. IIED.

IAASTD. (2009). *Agriculture at a crossroads: Synthesis report*. UNEP.

Illich, I. (2005). *Le silence fait partie des communaux. Le Miroir du passé. Oeuvres Complètes* (Vol. 2). Fayard.

Isenhour, C., & Reno, J. (2019). On materiality and meaning: Ethnographic engagements with reuse, repair & care. *Worldwide Waste: Journal of Interdisciplinary Studies, 2*(1), 1–8. https://doi.org/10.5334/wwwj.27

Jones, A., Pimbert, M. P., & Jiggins, J. (2012). *Virtuous circles: Values, systems, sustainability*. IIED and IUCN CEESP.

Kothari, A., Demaria, F., & Acosta, A. (2014). Buen Vivir, degrowth and ecological Swaraj: Alternatives to sustainable development and the green economy. *Development, 57*(3–4), 362–375. https://doi.org/10.1057/dev.2015.24

Kothari, A., Salleh, A., Escobar, A., Demaria, F., & Acosta, A. (2019). *Pluriverse: A post-development dictionary*. Tulika Books.

Kropotkin, P. A. (1898, revised 1913). *Fields, factories, and workshops, or industry combined with agricultures and brain work with manual work*. Benjamin Blom.

La Vía Campesina. (2008). *Declaration of Maputo: V International Conference of La Vía Campesina*. Retrieved July 19, 2021, from https://viacampesina.org/en/declaration-of-maputo-v-international-conference-of-la-via-campesina/

La Vía Campesina. (2020). *The United Nations declaration on the rights of peasants and other people working in rural areas*. La Via Campesina. Retrieved July 19, 2021, from https://viacampesina.org/en/the-united-nations-declaration-on-the-rights-of-peasants-a-tool-in-the-struggle-for-our-common-future/

Latouche, S. (2011). *Vers une société d'abondance frugale: contresens et controverses sur la décroissance* [Towards a society of frugal abundance: Misinterpretations and controversies about de-growth]. Fayard/Mille et une nuits.

Levidow, L., Pimbert, M. P., & Vanloqueren, G. (2014). Agroecological research: Conforming—or transforming the dominant agro-food regime? *Agroecology and Sustainable Food Systems, 38*(10), 1127–1155. https://doi.org/10.1080/21683565.2014.951459

Lilia, D., Espinal, T., & Peña Azcona, I. (2020). Care ethics in agroecology research: Practices from southern Mexico. *Farming Matters, 24*–27. Retrieved July 19, 2021, from http://www.cultivatecollective.org/in-practice/farming-matters-agroecology-feminism/

Mies, M., & Bennholdt Thomsen, V. (1999). *The subsistence perspective: Beyond the globalised economy.* Zed Books.

Milgroom, J. (2021). Linking food and feminisms: Learning from decolonial movement. In *Agroecology in Motion.* Retrieved July 8, 2021, from https://www.agroecologynow.com/linking-food-and-feminisms/

Norgaard, R. B., & Sikor, T. O. (1995). The methodology and practice of agroecology. In M. A. Altieri (Ed.), *Agroecology: The science of sustainable agriculture.* Westview Press.

Nyéléni (2007). *Declaration of the forum for food sovereignty, Nyéléni.* Retrieved December 23, 2015, from https://nyeleni.org/spip.php?article290

Öcalan, A. (2011). *Democratic confederalism.* Transmedia Publishing.

Perfecto, I., & Vandermeer, J. (2017). A landscape approach to integrating food production and conservation. In I. J. Gordon, H. H. T. Prins & G. R. Squire (Eds.), *Food production and nature conservation: Conflicts and solutions.* Routledge.

Pimbert, M. P. (2008). *Towards food sovereignty: Reclaiming autonomous food systems.* IIED and RCC. Retrieved July 20, 2021, from http://www.environmentandsociety.org/mml/towards-food-sovereignty-reclaiming-autonomous-food-systems

Pimbert, M. P. (2012). *Fair and sustainable food systems: From vicious cycles to virtuous circles.* IIED Policy Brief, International Institute for Environment and Development, London.

Pimbert, M. P. (2018a). *Food sovereignty, agroecology, and bio-cultural diversity: Constructing and contesting knowledge.* Routledge.

Pimbert, M. P. (2018b). Global status of agroecology, a perspective on current practices, potential and challenges. *Economic and Political Weekly, 53*(41), 52–57.

Pimbert, M. P. (2019). Food sovereignty. In P. Ferranti, E. M. Berry, & J. R. Anderson (Eds.), *Encyclopedia of food security and sustainability* (Vol. 3, pp. 181–189). Elsevier.

Pimbert, M. P. (2021). Regenerating Kurdish ecologies through food sovereignty, agroecology and economies of care. In S. Hunt (Ed.), *Ecological solidarity in the Kurdish freedom movement*. Lexington Books.

Pimbert, M. P., & Borrini-Feyerabend, G. (2019). *Nourishing life: Territories of life and food sovereignty*. Policy Brief of the ICCA Consortium no. 6. Teheran: ICCA Consortium, Centre for Agroecology, Water and Resilience at Coventry University and CENESTA. Retrieved September 18, 2020, from https://www.coventry.ac.uk/globalassets/media/documents/research-doc uments/research-projects/consortium-policy-brief-6-territories-of-life-and-food-sovereignty.pdf

Pimbert, M. P., Moeller, N. I., Singh, J., & Anderson, C. A. (2021). Agroecology. In M. Aldenderfer (Ed.), *Oxford research encyclopaedia of anthropology*. Oxford University Press. https://doi.org/10.1093/acrefore/978019 0854584.013.298

Polanyi, K. (1957). *The great transformation*. Beacon Press.

Pushpakumara, D., Marambe, B., Silva, P., Weerahewa, J., & Punyawardena, B. (2012). A review of research on homegardens in Sri lanka: The status, importance and future perspective. *Tropical Agriculturist, 160*, 55–125.

Rist, G. (2011). *The delusions of economics: The misguided certainties of a hazardous science*. Zed Books.

Rist, G. (2013). *Le développement: histoire d'une croyance occidentale* [Development: History of a western belief] (4th revised ed.). Presses de Sciences Po.

Ruivenkamp, G., & Hilton, A. (2017). *Perspectives on communing: Autonomist principles and practices*. Zed Books.

Salomeyesudas, B., & Satheesh, P. V. (2009). Traditional food system of Dalit in Zaheerabad Region, Medak District, Andhra Pradesh, India. In H. V. Kuhnlein et al. (Eds.), *Indigenous peoples' food systems: The many dimensions of culture, diversity and environment for nutrition and health*. FAO.

Sievers-Glotzbach, S., Tschersich, J., Gmeiner, N., Kliem, L., & Ficiciyan, A. (2020). Diverse seeds—shared practices: Conceptualizing seed commons. *International Journal of the Commons, 14*(1), 418–438. https://doi.org/10. 5334/ijc.1043

Sievers-Glotzbach, S., Euler, J., Frison, C., Gmeiner, N., Kliem, L., Mazé, A., & Tschersich, J. (2021). Beyond the material: Knowledge aspects in seed commoning. *Agriculture and Human Values, 38*, 509–524. https://doi.org/ 10.1007/s10460-020-10167-w

Steffen, W., Richardson, K., Rockström, J., Cornell, S. E., Fetzer, I., Bennett, E. M. et al. (2015). Planetary boundaries: Guiding human development on a changing planet. *Science, 347*(6223), 1–15. https://doi.org/10.1126/sci ence.1259855

Tittonell, P. (2020). Assessing resilience and adaptability in agroecological transitions. *Agricultural Systems, 184*, 102862.

UNFSS. (2021). *Policy options for food systems transformation in Africa—from the perspective of African universities and think tanks*. Food Systems Summit Brief Prepared by Research Partners of the Scientific Group for the Food Systems Summit. Retrieved July 21, 2021, from https://sc-fss2021.org/wp-content/uploads/2021/06/FSS_Brief_Policy_Options_Africa.pdf

UNDROP. (2019). *The United Nations declaration on the rights of peasants and other people working in rural areas*. United Nations. Retrieved July 19, 2021, from https://digitallibrary.un.org/record/1650694

UNEP. (2020). *Preventing the next pandemic: Zoonotic diseases and how to break the chain of transmission*. UNEP.

WEF. (2018). *Innovation with a purpose: The role of technology innovation in accelerating food systems transformation*. World Economic Forum.

White, R. J., & Williams, C. C. (2014). Anarchist economic practices in a 'capitalist' society: Some implications for organisation and the future of work. *Ephemera: Theory & Politics in Organization, 14*(4), 951–975.

Wright, J. (Ed.). (2021). *Subtle agroecologies: Farming with the hidden half of nature*. CRC Press.

CHAPTER 3

Integration of Endogenous Development Theory into the Study of Seed Governance

Yoshiaki Nishikawa

Abstract In the 1970s, the Japanese sociologist Kazuko Tsurumi developed endogenous development theory—the idea of 'development from within', which frames human wellbeing, ecological viability and community agency as central to sustainable modernisation. In this study of Tsurumi's ideas vis-à-vis seed governance, Yoshiaki Nishikawa first traces the broader debate over seed systems, from polarised stances such as traditional vs modern to more nuanced mixed approaches. Nishikawa shows how Tsurumi's thinking on values, communication, local autonomy and tradition can illuminate understanding of humanity's relationship with seeds across cultures and regions. Many farmers, for instance, consider crop diversity and seed production as naturally integral to their stewardship of local ecologies, rather than politicised acts of sovereignty. Wise governance is based on an understanding of seeds as a biocultural legacy, and ensures that autonomy and respect are interwoven in the concept and practice of seed sovereignty.

Y. Nishikawa (✉)
Ryukoku University, Kyoto, Japan
e-mail: nishikawa@econ.ryukoku.ac.jp

Keywords Endogenous development · Kazuko Tsurumi · Nara Document of Authenticity · Seed governance · Sustainable livelihoods approach

3.1   INTRODUCTION

This book has two objectives. One is to expand the understanding and application of agroecology and food sovereignty concepts, with an emphasis on the Asian context. The other is to render debates on mainstreaming agroecology and food sovereignty more diverse and inclusive.

Along with flourishing ecosystems and inter-species diversity, crop biodiversity is central to agroecology, and crucial in ensuring the sustainability of crop farming and food production. Diversity in crop species has been created by farmers themselves, especially those engaged in so-called traditional agriculture as practised in developing countries (Harlan, 1992). In this way, farmers directly contribute to the UN Sustainable Development Goal 15, 'Life on Land', which aims to protect, restore and promote the sustainable use of biodiversity.

This chapter introduces endogenous development theory, originally developed by a Japanese sociologist Kazuko Tsurumi in the 1970s and other researchers, mainly during the 1980s and 1990s, to help develop thinking on agroecology and sovereignty vis-à-vis the realities of rural communities, especially in terms of diversity and inclusion (see Chapter 1, Box 1.2). Focusing on the role of autonomous actors in resource management, and more generally on the importance of achieving a more harmonious relationship with nature, the chapter evaluates seed systems in which livelihoods are rooted in understanding of the surrounding environment and of humanity as part of it.

3.2   A BRIEF HISTORY
OF THE DEBATES ON SEED SYSTEMS

Crop-breeding institutions—most of which focus on biological science and the industrialisation of agriculture—are mainly concerned with obtaining sufficient diversity in terms of breeding materials. Their aim with conservation and the management of seeds as crop genetic resources is primarily utilitarian for future agriculture, largely ignoring current social and cultural dimensions (Ford-Lloyd & Jackson, 1986; Frankel & Soule, 1981; Srinivasan, 2010). Under this agenda, they promote the idea that genetic resources are a common heritage allowing everyone free access, even as they largely ignore the contribution of Indigenous people who developed and maintained crop biodiversity (Mooney, 1983).

These differing conceptions of crop diversity and its implications have led to conflicts between groups that represent farmers' and plant breeder's rights (Kloppenburg & Kleinman, 1988). Thus early on, many crop scientists (Frankel, 1988; Kawano, 2003) were proposing multidisciplinary approaches involving, for example, a combination of research on biological evolution with research on the social implications of diversity management.

Seed systems are often divided into formal and informal or local systems (Almekinders & Louwaars, 1994). A seed system encompasses a range of activities, including production, saving, distribution, certification and sales, as well as a variety of institutions that support these activities. An informal system mainly deals with the supply of seeds derived from non-certified local or traditional (open-pollinated) varieties that have been created through seed production or exchange by farmers. A formal system aims to provide seed certification, improve varieties (hybrids) and ensure supply under the control of a government body.

Thus, in traditional agriculture, farmers continue to engage in seed-saving practices (McGuire & Sperling, 2016). In industrialised agriculture, breeding and seed supply are mainly undertaken by governments and/or seed corporations, and farmers are generally obliged to purchase seeds every year.

Many governments have promoted seed-provision systems of the latter kind, based on 'scientific' evidence of more efficient production (Biggs, 2008). Yet in local systems, crop varieties are developed under specific regional climatic and soil conditions; their genotypes are particularly suited to these conditions, and not necessarily to those of other regions. Such varieties tend also to be woven into local culture, customs and livelihoods in ways that other varieties within the same species cannot replicate (Suge, 1987). Additionally, numerous farmers and hobby gardeners continue to collect seeds themselves in both developing and developed countries, including Japan (see Chapters 4–6).

3.3   The Diversification of Debates on Seed Systems and Governance

Perspectives on seed governance among various actors, including researchers, differ widely and often conflict: traditional vs modern, subsistence vs commercial, local vs global, small- vs large-scale frameworks. The failure of the current food and agriculture system has made these conflicts unavoidable (Duncan et al., 2019; Rosset & Altieri, 2017). However, strict application of any one framework may hinder the development of more sustainable and resilient systems unless attention is paid to the diverse mechanisms supporting seed resilience in each community. Both global and local approaches are important, for instance, but need to be evaluated differently.

Seeds as commons is a concept advocated by many seed activists, including Vandana Shiva (2020). Globally, such a commons demands the inclusion of diverse stakeholders—not only traditional players from public and private institutions, but also members of grassroots organisations, and farmers themselves. The concept is based on the framework of the 2001 International Treaty on Plant Genetic Resources for Food and Agriculture, or ITPGRFA, which promotes plant genetic resource conservation and protects farmers' rights to a fair share of any benefits arising from their use (Frison, 2018).

Indeed, many studies link farmers' rights, seed and food sovereignty, and 'right to food' frameworks to grassroots seed-saving activities, as explained in Chapter 2. The ITPGRFA clearly recognises the concept of farmers' rights as a counterweight to excessive protection of breeders' rights (Andersen, 2008; Esquinas-Alcazar et al., 2013). Achieving collaboration and harmonisation across different activities among diverse stakeholders is critical to ensuring sustainable, resilient seed provision and procurement (Nishikawa, 1990). A diversity of stakeholders is inevitable, given the range of different functions and facilities needed if both conservation and sustainable use are to succeed. Research must therefore encompass more than seed provision and procurement mechanisms, and involve stakeholders working together with others, especially farmers—the most important group in the context of seeds (Chambers, 2005; Neef & Neubert, 2011; Scoones, 2015).

There are many kinds of seed sourcing that cannot be simply explained in terms of single frameworks of seed governance as described above, such as traditional or modern. (These will be explored in subsequent chapters.)

Even when formal systems predominate, many farmers still use seeds originating in both informal and formal systems and obtain through purchase, self-saving or exchange (Coomes et al., 2015). And some private companies are in close contact with peasant farmers, enabling them to tailor seed sales to varieties suitable for their farms. By using these perspectives to evaluate seed systems in rural Japan, elsewhere in Asia and beyond—where livelihoods are deeply rooted in the understanding of surrounding nature and humankind as a part of it—we can add nuance to conceptualisations of agroecology and food sovereignty and enable more resilient seed governance.

As explained in Chapter 2, policies that apply the tenets of agroecology and sovereignty have ultimately helped in transforming highly industrialised, globalised and increasingly unsustainable and fragile food and agriculture systems into more sustainable and resilient ones (Anderson et al., 2020; Levidow et al., 2014; Pimbert, 2018). Jessica Duncan and colleagues (2019), in promoting agroecology research, state: 'We continue to think deeply about our theories of change, our values and principles, and the different roles that we can and do play in enabling agroecology transformations in the food sovereignty movement at large'. Colin Anderson and colleagues (2020) meanwhile assert that bottom-up forms of governance and the self-organisation of communities have the potential to enable transformation for sustainability. Various forms of seed activism also exist in Europe, South America and other regions (Peschard & Randeria, 2020).

Anderson and colleagues suggest that approaches that co-opt or disable bottom-up processes, which are often deployed by governments or corporations, should be avoided. As the core ideas of agroecology and food sovereignty reflect the principles of endogenous development, those principles inevitably enrich any discussion of agroecological transformation. However, endogenous development theory clearly embraces the coexistence of a diversity of systems; and if any model or frame, such as agroecology, is to be widely applied, it needs to be done carefully, given the range of conditions across locales in both Asian and Western contexts.

3.4   Endogenous Development Theory: A 'Third System' for Understanding Development

The concept of endogenous development was first identified in the report *Development and International Cooperation*, which was submitted by the Hammarskjöld Foundation to the UN General Assembly in 1975 (Dag Hammarskjöld Foundation, 1975). The report states: 'If the goal of development is human development that aims at the liberation and self-development of man as an individual or as a social being, that development must be endogenous in each social community'.

At around the same time, Kazuko Tsurumi began to apply the concept of endogenous development to specific cases in China and Japan. One of her foci was communities affected by Minamata disease, a severe neurological disease caused by methylmercury poisoning that was first discovered in Minamata, Japan, in 1956 and linked to industrial waste (Tsurumi, 1989, 1996). As one of the foremost proponents of endogenous development theory, Tsurumi set out to define its key elements.

First, endogenous development is specifically associated with certain values, in direct contrast to the standardised or generalised process of modernisation theory, which tends to take a neutral stance on values. (The German sociologist Max Weber was among the first to theoretically codify the transition from traditional to modern society.) Tsurumi states that whereas modernisation theory tends to generalise the process of transition regardless of historical, social and cultural distinctions, endogenous theory focuses on specific societies, is less abstract and is more oriented towards local communities. While any approach applying endogenous development has a common goal—achieving wellbeing—Tsurumi's formulation posits a number of ways to achieve that, and a range of societal models that might be more likely to enable these processes, as well as other changes that might ensue, such as emergence of different key persons for development.

Second, endogenous development theory emphasises the importance and possibility of mutual exchange. Although endogenous development takes place simultaneously in many localities on a global scale, the kind of development varies; and further, each kind can become a model for other forms of development. In the 1970s, for example, people in Minamata exchanged experiences with Indigenous communities in Canada who had also been affected by mercury poisoning. Despite living under completely different economic, social, cultural and political conditions, the two groups shared ideas how corporate interests and government pressure had severely undermined their dignity and created health crisis and community collapse. This exchange led to involvement of physicians to refine the diagnostics.

The third element Tsurumi isolated is that a region is an ideal unit for endogenous development, at least in theory. It is a smaller unit than a nation, but not necessarily contained within one nation—such as the Basque Country, which straddles the border of Spain and France.

Fourth, endogenous development rests on a 'third system' of power, distinguishing it from other development theories, which tend to refer to political power as the most important, followed by economic power. This third system focuses not on seeking to change existing political and economic power structures, but rather to encourage locals to have their say and act autonomously within them.

Finally, Tsurumi stressed the importance of tradition. That is, institutions and norms well accepted by people within the designated regions, and comprising three aspects: a structure of awareness, social relations and practical techniques that express the dynamic nature of tradition.

In understanding Tsurumi's conceptualisation of endogenous development, her experience in Minamata reveals much. She visited four communities in the city where many had suffered from Minamata disease. She and her group conducted interviews with study group members, rather than using questionnaires or quantitative data collection methods. From their findings, Tsurumi drew two key conclusions (Tsurumi, 1989, 1996). One was that Minamata disease provided the most extreme case of certain negative aspects of centralised and corporate-led modernisation; the other was that in terms of endogenous development, the actors were people who had suffered from those negative impacts, enacted by the government and certain enterprises.

Tsurumi also developed the concept of autonomous rehabilitation. In the case of Minamata disease, the destruction of nature through rampant industrialisation and pollution affected people not only externally—that is, in their relationship with nature and with other people. It also had an internal impact, on their physical and spiritual health and dignity. Autonomous rehabilitation enabled healing in both respects. When people with the disease realised the limitations of modern medicine in tackling the damage they had suffered, they decided to rehabilitate themselves by deploying their own capabilities and creativity (see Box 3.1). Tsurumi's view of society as multi-faceted posits that transformation cannot be unilateral; that in turn provides a stimulus for finding diverse and inclusive ways of developing relationships between nature and humanity in relation to seed governance.

Box 3.1: Kazuko Tsurumi's Case Study in Minamata: The Findings
Towards the end of the 1970s, Tsurumi selected three people with differing attitudes towards nature, and towards the institutions implicated in the Minamata disaster, to examine their personal rehabilitation and that of the Minamata community. She identified the approaches observed in these three individuals as confrontational-integrative, non-confrontational fusionist, and non-confrontational and integrative.

- The actor whose approach was confrontational-integrative organised a self-help group of people with Minamata disease, and filed lawsuits against the company involved, the local government and eventually the national government. He also asserted the necessity of engaging in direct negotiation with the company's president.
- The actor favouring the non-confrontational fusionist approach focused on healing and rehabilitation by immersing in nature, specifically through local activities such as fishing, as well as engaging more closely with her neighbours.
- The non-confrontational integrative actor initially mobilised actively against the company and the public authorities involved. After attending a UN Human Environmental Conference in Stockholm in his wheelchair, however, he became aware of other people affected by mercury poisoning—in countries such as Canada and Indonesia—and began to communicate with them as co-sufferers.

Tsurumi described these creative approaches as demonstrating three layers of social change, seeking a return to an idealised past, cure and rehabilitation focused on both nature and self, and renovative creation of balance in which conflicting ideas coexist simultaneously (Tsurumi, 1987).
Source Originally Presented at the Ninth Meeting of the International Society for the Study of Behavioral Development, 13 July 1987, Tokyo.

There are other definitions and applications of endogenous development theory. One core point of debate has been whether endogenous development can be integrated into government policy, which is generally perceived as the imposition of edicts or measures on communities by an external institution. In such cases, the argument goes, endogenous development processes could no longer be considered properly endogenous, and would generate unresolvable tension between enforceable policy and local people's spontaneous judgement or practice (Matsumiya, 2001), which is independent of influence from outside with power. If authority is exercised uniformly over all communities and/or individuals, it could be seen as conflicting with local autonomy.

Locally specific issues arising in different localities need constant re-evaluation, and endogenous development should be considered as one potentially valuable tool in understanding how conflicting practices and theories can coexist.

In relation to endogenous development, Hisashi Nakamura (1989) theorised that agriculture can be seen as a symbiotic relationship between humans and non-humans, and is thus tied to the specific space and time in which symbiosis is created and maintained. In this book, especially Chapters 5, 7, and 10 about Japan and East Asia, contributors elaborate on this relationship to expand current understanding of agroecology by integrating it with components of endogenous development theory.

Readers of this book seeking to understand endogenous development better should note that relevant views have been theorised in the sustainable livelihoods approach (SLA), which was widely adopted by the former UK Department for International Development from the 1990s (Scoones, 2015). The SLA emphasises diversity in its analysis of livelihoods. In this it is assumed that, while those from outside a community may provide assistance, it is local people who are trying to diversify their lifestyles by applying a strategy prioritising sustainable survival.

In the SLA, the starting point for development, or resource governance, is that such governance is sustainable only when rural people spontaneously participate in it, rather than be compelled to do so by external institutions. Indeed, as Ian Scoones has noted (2015), the 'failure of agricultural and rural development to date has been largely due to the uniform top-down policies of governments and aid agencies (focused on economic growth)'. Although the rationale is different, the SLA is similar to endogenous development theory in according a lesser role to outside agencies in community development than to the communities themselves.

3.5   Approaches to Exploring Seed Governance and Expanding Perspectives on Seed Sovereignty

Farmers are key players in conserving biocultural diversity—that is, the integration of farming into people's livelihoods. In developing countries, many farmers do not necessarily prioritise increased crop yields or profits; they also consider cultural values, avoidance of risks (such as drastic harvest failures), and personal preferences in crop varieties. Long before the concept of seed sovereignty arose within political debates, farmers and local people tended to see crop diversity and seed production as just aspects of their everyday work.

Moreover, not all people who procure and supply seeds in Japan and other Asian countries base their farming practice or activities on specific rights to use and control resources. Their approaches and practices are derived not from politics but from a kind of intuitive understanding, as farmers, of their role in stewarding regional ecology. That process expresses their own endogenous development, and provides the foundation of their approach to seed care. For many of them, there is no contradiction between concepts of property rights and informal, day-to-day applications of traditional tacit knowledge on seeds: the two can exist in parallel (Rival, 2018; Nakazora, 2019).

Each chapter of this book looks at the approaches of ordinary people who do not think in terms of sovereignty over seeds. Sam Gray and Raj Patel (2015), among others, have claimed that ideas of food sovereignty have the potential to liberate people and communities that are otherwise deeply dominated by unsustainable food and agriculture systems. However, introducing the concept of seed sovereignty to seed governance where people have emphasised commitment and care rather than rights-based thinking may run the risk of generating new forms of oppression. A sovereignty-based approach may inadvertently pressure people who have managed their resources without thinking in such terms to accept the political perspective of sovereignty as necessary discourse (Anderson et al., 2020).

All the contributors of this book seek to encourage a greater awareness and acceptance of agroecology and food sovereignty. At the same time, they honour the fact that many people hesitate to introduce such ideas into local communities and institutions without first understanding and safeguarding local stakeholders' evolved approaches to governing resources.

These concerns can be illustrated by debates that took place in conferences of the International Council on Monuments and Sites (ICOMOS) during the 1990s, where delegates and members wrangled with the concept of authenticity vis-à-vis cultural heritage and cultural diversity (Stovel, 2008). The result was an important declaration, the Nara Document on Authenticity. Before 1992, when Japan became a member of the World Heritage Convention (which had been adopted by the General Conference of the United Nations Educational, Scientific and

Cultural Organization, or UNESCO, in 1972), the issue of heritage authenticity was mainly discussed in the context of European stone structures. Wooden structures that are periodically rebuilt based on traditional designs and methods—such as the Shinto Ise Grand Shrine in Japan—were not recognised as part of authentic heritage. (Since each of more than 700 thousand pieces of construction materials need to be examined for the authenticity every twenty years under UNESCO scheme, the Shrine decided not to be registered as a World Heritage. However, another site, Shirakawa, with wooden houses whose materials for thatches are gathered in the traditional way was registered because it is the way itself that is recognised as authentic.)

The Nara Document on Authenticity (ICOMOS, 1994) affirms that cultural heritage diversity exists across time and space and that other cultures and all aspects of their belief systems need to be respected (Article 6). Specifically, Article 11 states:

> *All judgments about values attributed to cultural properties as well as the credibility of related information sources may differ from culture to culture, and even within the same culture. It is thus not possible to base judgments of values and authenticity within fixed criteria. On the contrary, the respect due to all cultures requires that heritage properties must be considered and judged within the cultural contexts to which they belong.*

Seeds are tangible substances forming an essential part of biocultural heritage and diversity within a sustainable seed governance system, which with food sovereignty forms an integral component of a sustainable and resilient society. UNESCO clearly recognises the importance of specific contexts in time and space, and of values recognised and asserted by parties concerned, while accepting the universal value of world heritage. Thus, when applying evaluation criteria and processes developed by outsiders, it is necessary to ensure meaningful debates through the inclusion of relevant stakeholders.

Seeking to promote one ideology from one specific society as authoritative is antithetical to diversity and inclusion. This book seeks to promote those values, which also lie at the core of the UN Sustainable Development Goals. Each chapter provides valuable information on the actors, processes and methods that are currently involved in the conservation and utilisation of crop genetic resources, and on the diversity of seeds in different contexts. These cases illustrate the mechanisms needed to prevent new forms of oppression directed towards local stakeholders, and how autonomy and respect can be secured as critical components of sovereignty.

References

Almekinders, C., & Louwaars, N. (1994). Local seed systems and their importance for an improved seed supply in developing countries. *Euphytica, 78*(3), 207–216. https://doi.org/10.1007/BF00027519

Andersen, R. (2008). The international treaty on plant genetic resources for food and agriculture with the international undertaking on plant genetic resources. In R. Andersen (Ed.), *Governing agrobiodiversity: Plant genetics and developing countries* (pp. 87–115). Ashgate.

Anderson, C. R., Bruil, J., Chappell, M. J., Kiss, C., & Pimbert, M. P. (2020). From transition to domains of transformation: Getting to sustainable and just food systems through agroecology. *Sustainability, 11*(19), 1–28. https://doi.org/10.3390/su11195272

Biggs, S. (2008). The lost 1990s? Personal reflections on a history of participatory technology development. *Development in Practice, 18*(4–5), 489–505.

Chambers, R. (2005). *Ideas for development*. Routledge.

Coomes, O. T., McGuire, S. J., Garine, E., Caillon, S., McKey, D., Demeulenaere, E., Jarvis, D., Aistara, G., Barnaud, A., Clouvel, P., Emperaire, L., Louafi, S., Martin, P., Massol, F., Pautasso, M., Violon, C., & Wencélius, J. (2015). Farmer seed networks make a limited contribution to agriculture? Four common misconceptions. *Food Policy, 56*, 41–50. https://doi.org/10.1016/j.foodpol.2015.07.008

Dag Hammarskjöld Foundation. (1975). *The 1975 Dag Hammarskjöld Report on Development and International Cooperation*. Dag Hammarskjöld Foundation. Retrieved July 26, 2021, from www.daghammarskjold.se/publication/1975-dag-hammarskjold-report-development-international-cooperation/

Duncan, J., Claeys, P., Rivera-Ferre, M. G., Oteros-Rozas, E., Van Dyck, B., Plank, C., & Desmarais, A. A. (2019). Scholar-activists in an expanding European food sovereignty movement. *The Journal of Peasant Studies*, 1–26. https://doi.org/10.1080/03066150.2019.1675646

Esquinas-Alcazar, J., Hilmi, A., & Noriega, I. (2013). A brief history of the negotiation on the International Treaty on Plant Genetic Resources for Food and Agriculture. In M. Halewood, I. Noriega, I., & S. Louafi (Eds.), *Crop genetic resources as a global commons* (pp. 135–149). Routledge.

FAO. (1996). *Report on the state of the world's plant genetic resources for food and agriculture*. FAO.

Ford-Lloyd, B., & Jackson, M. (1986). *Plant genetic resources: An introduction to their conservation and use*. Edward Arnold Ltd.

Frankel, O. H. (1988). Genetic resources: Evolution and social responsibilities. In J. K. Kloppenburg (Ed.), *Seeds and sovereignty: The use and control of plant genetic resources* (pp. 19–46). Duke University Press.

Frankel, O. H., & Soule, M. E. (1981). *Conservation and evolution.* Cambridge University Press.

Frison, C. (2018). *Redesigning the global seed commons: Land and policy for agrobiodiversity and food security.* Routledge.

Grey, S., & Patel, R. (2015). Food sovereignty as decolonization: Some contributions from Indigenous movements to food system and development politics. *Agriculture and Human Values, 32,* 431–444. https://doi.org/10.1007/s10460-014-9548-9

Harlan, J. R. (1992). *Crops and Man* (2nd ed.). American Society of Agronomy-Crop Science Society. ISBN 978-0891181071.

ICOMOS. (1994). *The NARA document on authenticity.* Retrieved July 26, 2021, from https://www.icomos.org/en/charters-and-texts/179-articles-en-francais/ressources/charters-and-standards/386-the-nara-document-on-authenticity-1994

Kawano, K. (2003). Thirty years of cassava breeding for productivity: Biological and social factors for success. *Crop Science, 43,* 1325–1335.

Kloppenburg, J., & Kleinman, D. (1988). Plant genetic resources: The common bowl. In J. K. Kloppenburg (Ed.), *Seeds and sovereignty: The use and control of plant genetic resources.* Duke University Press.

Levidow, L., Pimbert, M. P., & Vanloqueren, G. (2014). Agroecological research: Conforming—or transforming the dominant agro-food regime? *Agroecology and Sustainable Food Systems, 38*(10), 1127–1155. https://doi.org/10.1080/21683565.2014.951459

Matsumiya, A. (2001). 'Naihatsuteki hatten' wo meguru shomondai, Naihatsuteki hattenron no tenkai ni mukete [Problems dealing with the concept of endogenous development: Searching for an alternative strategy in studies of endogenous development]. *Social Welfare Studies, 3*(1), 45–54.

McGuire, S., & Sperling, L. (2016). Seed systems smallholder farmers use. *Food Security, 8,* 179–195. https://doi.org/10.1007/s12571-015-0528-8

Mooney, P. (1983). *The law of the seed: Another development and plant genetic resources.* Development Dialogue series, 1–2. Dag Hammarskjöld Foundation.

Nakamura, H. (1989). Technology and autonomous local development movement. In K. Tsurumi & T. Kawata (Eds.), *Endogenous development theory* (pp. 215–240). University of Tokyo Press.

Nakazora, M. (2019). Temporalities in translation: The making and unmaking of "folk" Ayurveda and bio-cultural diversity. In K. Omura, J. G. Othuki,

S. Satsuka, & A. Morita (Eds.), *The world multiple: The quotidian politics of knowing and generating entangled worlds* (pp. 140–154). Routledge.

Neef, A., & Neubert, D. (2011). Stakeholder participation in agricultural research projects: A conceptual framework for reflection and decision-making. *Agriculture and Human Values, 28,* 179–194. https://doi.org/10.1007/s10 460-010-9272-z

Nishikawa, Y. (1990). *Institution for plant genetic resources in developing countries.* Papers in Development Administration No. 37. University of Birmingham.

Peschard, K., & Randeria, S. (2020). 'Keeping seeds in our hands': The rise of seed activism. *The Journal of Peasant Studies, 47*(4), 613–647. https://doi.org/10.1080/03066150.2020.1753705

Pimbert, M. (2018). Democratizing knowledge and ways of knowing for food sovereignty, agroecology and biocultural diversity. In M. Pimbert (Ed.), *Food sovereignty, agroecology and biocultural diversity: Constructing and contesting knowledge* (pp. 259–321). Routledge.

Rival, L. (2018). An anthropological lens on property and access: Gudeman's dialectics of community and market. In F. Girald & C. Frison (Eds.), *The commons, plant breeding and agricultural research: Challenges for food security and agrobiodiversity* (pp. 147–158). Routledge.

Rosset, P. M., & Altieri, M. A. (2017). *Agroecology: Science and politics.* Fernwood Publishing.

Scoones, I. (2015). *Sustainable livelihoods and rural development.* Agrarian Change & Peasant Studies Series. Fernwood Publishing.

Shiva, V. (2020). *Reclaiming the commons; biodiversity, indigenous knowledge, and the rights to Mother Earth.* Synergetic Press.

Srinivasan, C. S. (2010). Plant breeders' rights and on-farm conservation. In S. Lockie & D. Carpenter (Eds.), *Agriculture, biodiversity and markets: Livelihoods and agroecology in comparative perspective* (pp. 61–76). Earthscan.

Stovel, H. (2008). Origins and influence of the Nara document on authenticity. *APT Bulletin, 39*(2/3), 9–17.

Suge, H. (1987). *Ikushu no genten* [Real base of plant breeding]. Nosanhyoson Bunka Kyokai.

Tsurumi, K. (1987). New lives: Some case studies in Minamata. In K. Tsurumi (Eds.), *The adventure of ideas: A collection of essays on patterns of creativity & a theory of endogenous development.* Japanime, Manga University. Retrieved July 26, 2021, from www.howtodrawmanga.com/products/tsurumi

Tsurumi, K. (1989). Naihatsu-teki Hatten-ron no keifu [A genealogy of endogenous development]. In T. Kawata & K. Tsurumi (Eds.), *Naihatsu-teki Hatten-ron* [Endogenous development]. Iwanami Shoten.

Tsurumi, K. (1996). Genealogy of endogenous development theory. In K. Tsurumi & T. Kawata (Eds.), *Endogenous development theory* (pp. 43–64). University of Tokyo Press.

CHAPTER 4

Indigenous Seed Systems and Biocultural Heritage: The Andean Potato Park's Approach to Seed Governance

Krystyna Swiderska and Alejandro Argumedo

Abstract In the Indigenous worldview, seeds are both biological entities and embodiments of immateriality: knowledge, culture and the sacred. Indigenous seed systems thus codify the human connection to nature. Yet such 'informal' systems, whether developed by Indigenous peoples or small-scale farmers, barely surface in policy debates. Krystyna Swiderska and Alejandro Argumedo seek to redress the balance in this detailed study of the principles, values and practices of Indigenous seed systems and governance. While ranging over a number of case studies from Kyrgyzstan to Kenya, their prime focus is the Andean Potato Park in

K. Swiderska (✉)
International Institute for Environment and Development (IIED), London, UK
e-mail: krystyna.swiderska@iied.org

A. Argumedo
Asociación ANDES (Association for Nature and Sustainable Development), Cusco, Peru

Swift Foundation, Santa Barbara, United States

© The Author(s) 2022
Y. Nishikawa and M. Pimbert (eds.), *Seeds for Diversity and Inclusion*,
https://doi.org/10.1007/978-3-030-89405-4_4

Cusco, Peru—a world centre of origin and domestication of crops such as the potato, quinoa and amaranth. Swiderska and Argumedo describe the Park's collective and customary governance structure, and the ways of learning, exchange systems, seed banks and more developed by its Quechua farmers. To safeguard the vital Indigenous contribution to seed security and diversity, they conclude, a biocultural rights-based approach to seed governance is required and needs further support from policy reform, among other measures.

Keywords Indigenous people · Biocultural heritage · Customary seed governance · Territories

4.1   INTRODUCTION

Indigenous peoples' seed systems sustain a rich diversity of underutilised species and varieties, both cultivated and wild, including ancestral populations of crops in centres of origin (i.e. domestication) (Bellon, 1996), based on Indigenous knowledge, values and worldviews (Graddy, 2013; Pilgrim & Pretty, 2010). Many Indigenous territories overlap geographically with centres of origin of crops, known as Vavilov centres (Maxted et al., 2020).

Seeds inevitably combine material and immaterial aspects such as knowledge and culture (Sievers-Glotzbach et al., 2021). This is very clearly seen in Indigenous seed systems. In Andean cultures, for example, seeds are regarded as spiritual beings connected to a landscape where everything, including rocks, has a spiritual dimension. Thus, seeds and seed systems are not only regarded as biological and economic resources, but also as socio-ecological systems governed by ancestral rules and values that emanate from Indigenous ways of understanding the universe, and that codify a deep respect for nature (Graddy, 2013).

Despite their critical importance for food and nutrition security and climate resilience, these so-called "informal" seed systems remain poorly understood (Gill et al., 2013). Thus, the distinct biocultural and normative character of Indigenous seed systems is often overlooked rather than supported by formal seed policies and conservation practices.

This chapter explores the customary principles, values and practices that characterise Indigenous seed systems as biocultural heritage, both globally and in the Andean Potato Park in Cusco, Peru—or Parque de la Papa—which enshrines a Quechua rights-based biocultural approach to seed governance. Here, the wild and the sacred realms play a greater role in seed governance than humans, and the holistic Andean worldview and values continue to play a critical role in ensuring biodiversity conservation, food security and human wellbeing. The chapter also explores the governance tools developed by Potato Park communities, with support from the NGO Asociación ANDES, to revitalise and conserve the Potato Park seed system as one of the world's richest in-situ genetic reserves.

4.2   Indigenous Seed Systems as Biocultural Heritage

Many Indigenous seed systems are guided by Indigenous core values of sharing and reciprocity, and balance with nature (Swiderska et al., 2009). These core values or principles are a common feature across Indigenous cultures from the Quechua of Peru to the Kuna of Panama, the Himalayan Lepchas, the Naxi of Yunnan, China, and the Mijikenda of coastal Kenya (Swiderska et al., 2011). The obligation to share seeds and related traditional knowledge enhances the diversity of seed each farmer holds, helps to maintain the purity of seed and promotes further diversification through adaptation to different environments (Argumedo et al., 2011). In the Indian Himalayas, for example, exchange of seeds between communities at different altitudes has enabled farmers to adapt to warmer climates (Pant, 2012), while in Southwest China, seeds are exchanged over very large distances (Swiderska et al., 2011). Women play a key role in managing seed systems and transmitting knowledge to younger generations (Swiderska et al., 2018).

In Cusco's Potato Park, Indigenous practices of saving and sharing seeds to spread risk are embedded in traditional networks that connect farmers across different environments (ANDES, 2016). The Potato Park communities have conserved a very high diversity of cultivated, semi-wild and wild crops, ensuring food security despite significant climate change impacts and the Covid-19 pandemic. This biodiversity exists thanks to their ancestral principles of solidarity, reciprocity and balance between humans, nature and the sacred worlds. As Mariano Sutta, a community expert from the Potato Park explained: "This diversity of food could not exist without these principles. For us having our food system built on these principles is very important" (Swiderska & Ryan, 2021).

For Indigenous peoples and small-scale farmers, seeds often have spiritual and ritualistic significance (Pilgrim & Pretty, 2010; Samuel & David, 2007). Seeds embody knowledge, practices and beliefs, inextricably linking biodiversity and intangible cultural heritage as "biocultural heritage" (see Box 4.1). In the Potato Park, seeds have souls, form communities, and have a system of rules that humans do not understand. When the International Potato Centre returned native potatoes that had been lost, this revived associated traditional knowledge and beliefs embedded in seeds (Swiderska et al., 2011). In Southwest China, restoring traditional seeds revived associated traditional knowledge and practices (Swiderska et al., 2009).

Even where spiritual beliefs have been weakened, the cultural values and uses of seeds—such as to produce traditional foods for ceremonies and festivals—often play a critical role in preventing the loss of genetic diversity (Swiderska et al., 2009). Across the world, Indigenous elders, and women in particular, continue to conserve traditional seeds and plant them in home gardens (African Biodiversity Network and Gaia Foundation, 2015; Swiderska et al., 2018).

Box 4.1: Indigenous biocultural heritage

The concept of biocultural heritage derives from Indigenous traditions and holistic worldviews. In 2005, Asociación ANDES, IIED and partners developed a definition of biocultural heritage based on decolonising research in the Potato Park, and the concept of Traditional Resources Rights (Argumedo & Pimbert, 2005; Posey et al., 1996). Their aim was to provide a common conceptual framework for research on protecting Indigenous knowledge.

They defined biocultural heritage as

"*Knowledge, innovations and practices of Indigenous peoples and local communities, that are collectively held and inextricably linked to traditional resources and territories, local economies, the diversity of genes, species and ecosystems, cultural and spiritual values, and customary laws, shaped within the socio-ecological context of communities*" (Swiderska et al., 2009).

Further participatory research with 11 Indigenous groups in the Potato Park Peru, Panama, India, China and Kenya confirmed these interlinkages and interdependencies, both in Indigenous worldviews and in practice. The research characterised biocultural heritage as a complex adaptive system with multiple interlinkages between its components (https://biocultural. iied.org).

Indigenous languages are key components of biocultural heritage as a means through which it is expressed and transmitted, along with memory, history, and ways of life within a particular territory and ecological context (Swiderska et al., 2020).

Crop wild relatives (CWRs) are crucially important in Indigenous farming traditions. Many Indigenous farmers still actively encourage the flow of resilient genes from wild to domesticated populations, and domesticate wild food plants to create new crops (Wilson, 2009). In Ethiopia, for example, farmers are finding and bringing into cultivation new strains of wild crops such as coffee and enset, a key root crop. And in Kyrgyzstan, farmers interplant wild apricot trees into their orchards to improve pollination (Wilson, 2009). In the Philippines, India and China, Indigenous farmers plant resilient CWRs to enrich domesticated crops in home gardens (Swiderska & Ryan, 2021). In Kenya, Mijikenda farmers continue to cultivate wild plants from sacred *kaya* forests on-farm (Swiderska et al., 2018).

Wild seeds, nuts, fruits, leaves, bush-meat and fish remain a significant micronutrient source for millions of Indigenous and rural people worldwide (Rowland et al., 2017). Wild food plants tend to be richer in micronutrients than cultivated crops, and more resilient (Borelli et al., 2020; Hunter et al., 2019). The role of Indigenous peoples in sustaining vital evolving gene pools and promoting gene flows to enhance resilience in domesticated crops is often overlooked in state policies. For example, member states participating in the FAO Commission on Genetic Resources for Food and Agriculture have advocated for separate networks for on-farm conservation by farmers, and for in-situ conservation of CWRs by states largely through existing protected areas.[1] Yet the majority of CWRs occur outside protected areas (Hunter & Heywood,

---

[1] http://www.fao.org/fileadmin/templates/agphome/documents/PGR/Reports/Report-Technical_workshop_131112.pdf.

2011), on the lands of Indigenous peoples in centres of crop diversity and domestication (Maxted et al., 2020).

Seed systems often lie at the heart of Indigenous peoples' struggles for self-determination and food sovereignty (Gutierrez, 2016). Maintaining the autonomy of such systems is vital to controlling farming and food decisions, resilience and cultural integrity (African Biodiversity Network and Gaia Foundation, 2015; AFSA & GRAIN, 2018). Many Indigenous peoples have, however, reported aggressive promotion of modern varieties that create dependence on costly external inputs (FAO, 2018).

Another key concern of Indigenous peoples is the privatisation of ancestral seeds through intellectual property rights (IPRs) such as patents and plant variety protection—a practice that goes against customary norms regarding collective custodianship and sharing of seeds to ensure access (Munyi & De Jonge, 2015; Swiderska et al., 2009). According to farmers in Ethiopia, "They [seeds] have personality. Farmers respect them as sacred gifts from nature, so that their seeds cannot be held in custody, privatised or patented by individuals: rather, seeds belong to the entire community" (AFSA & GRAIN, 2018).

African regional intellectual property organisations, representing several countries, have adopted Plant Variety Protection (PVP) protocols modelled on European standards of the International Union for the Protection of New Varieties of Plants (UPOV) Convention 1991. This poses a threat to farmers' rights and seed systems because UPOV'91 limits the possibilities for farmers to use, exchange and trade farm-saved seed of protected varieties (Munyi & De Jonge, 2015). IPRs, as well as non-IPR-related seed laws requiring certification and standardisation, are increasingly criminalising informal seed systems and restricting "seed commons" (Sievers-Glotzbach et al., 2021; Wattnem, 2016).

4.3   The Potato Park's Andean Seed System

The Andean region is one of the world's centres of origin and domestication of food crops including the potato and other tubers (olluco, oca), and the "superfoods" quinoa and amaranth (Sayre et al., 2017). Since 1998, Asociación ANDES has worked with six (now five) Quechua communities to establish the Potato Park.

This biocultural heritage territory in a secondary centre of origin of the potato (CIP, 2008) near Pisaq, Cusco, Peru, is collectively managed. Local communities joined their land to form an area of around 9600 hectares, and in 2002 legally registered a collective Potato Park Association. Guided by pre-colonial Andean cosmovision and customary principles, the Potato Park conserves rich biodiversity, including some 1300 native potato cultivars (or, based on Western classification, around 650 varieties), 4 potato wild relatives, and many other Andean tubers, grains, medicinal plants and wildlife (ANDES, 2016).

A single plot can contain 250–300 potato varieties (Jiggins, 2017). Food sovereignty and protection of Indigenous rights are key objectives. As Ricardina, a local expert at the Potato Park explained, the territory "is an example of how communities can come together to defend their land" (Swiderska & Ryan, 2021) (see Box 4.2).

Box 4.2: Governance of the potato park

The Potato Park's five communities, sitting at an altitude of 3400–4600 metres above sea level, are home to over 6000 Quechua people (ANDES, 2016). The park is collectively governed by a general council composed of the elected leaders of each community. The council oversees local technical experts who are elected by each community, including a Papa Arariwa or Potato Guardians collective, which supports the park's communal seed system. The Potato Park Association and Council have an administration centre to support the guardians and the park's economic collectives: Medicinal Plants (herbal teas, shampoos, creams), Gastronomy, Agro-ecotourism and Handicrafts (traditional weaving). These micro-enterprises bring together experts from different communities to produce a range of biocultural products and services (Aniceto Ccoyo, in Swiderska & INMIP, 2017).

Ten per cent of the revenues generated by each micro-enterprise is invested in a communal fund and redistributed annually among the communities in accordance with Andean principles of reciprocity and equilibrium that promote equity, ecological sustainability and solidarity. These principles are set out in an Inter-Community Agreement for Benefit Sharing, or Biocultural Heritage Protocol, which aside from ensuring equitable sharing of benefits provides the basis for collective governance of the park (ANDES et al., 2011). The agreement follows the principles embedded in Andean cosmovision—the Ayllu—where the wild and the sacred realms, embodied in the Apus or mountain Gods, play a key role in governance (Swiderska & INMIP, 2017).

The Apus "are the highest authority in the governance of the Potato Park", notes Mariano Sutta (community expert, Swiderska & Ryan, 2021).

In this context, Aniseto Ccoy, community expert of the Saccaca Community has noted: "It is their laws we have to follow because they are older than humans and because our customary laws are based on how these 'big heads' think. One mountain has been elected as 'mayor', the head of the whole land. The other Apus are councillors of water, medicinal plants and so on" (Swiderska & INMIP, 2017).

Potato Park farmers use only native seeds because "modern varieties don't produce well" in the territory (Potato Park expert, in Swiderska & Ryan, 2021). Native potatoes are not grown commercially, but for food, using traditional agroecological practices. While many landraces, or local cultivars, do not produce high yields compared to industrially farmed crops, their diversity ensures resilience to shocks, stable productivity and nutrition. The diversity of native potatoes is also a source of pride and a symbol of the maintenance of tradition in Andean communities (Sayre et al., 2017).

Different types of native potatoes have different cultural significance. Qachun Huaccachi—which translates as "makes the bride cry", as it is very difficult to peel—is considered a symbol of love (Swiderska & INMIP, 2017). Native potato seeds are traditionally gifted to young couples by both families, disseminating varieties and related history, stories and knowledge about their uses and special characteristics (Walshe & Argumedo, 2016). The tubers are nurtured by Quechua women as "children" (see Box 4.3). The Potato Park has revived a number of rituals and festivals linked to native potatoes, e.g. offering gifts to the spirit of the potato, and tying a rope around harvested potatoes to keep the spirit tied to the earth—"we have to keep the spirit inside so the potatoes are better and stronger" (Potato Park expert, Swiderska & INMIP, 2017) (see Fig. 4.1).

Box 4.3. Andean learning and nurturing values that shape seed governance

In the Andean model, there are three distinct kinds of learning or knowledge:

- *yachay* is knowledge or wisdom that is processed mentally and learned through reflection, discussion and analysis;

- *ruway* (or *llankay*) is practical learning, for instance of skills related to agriculture or food preparation;
- *munay* is emotional learning, signifying love between people, and between people and nature, and refers to social connections, intuition, desire and the capacity to think and feel with the heart (Swiderska & Stenner, 2020).

This way of understanding knowledge creation can be applied to seed systems, and is situated within a place-based context—the wisdom relates to the landscape and its components. Feeling and rituals related to people's deep respect for nature are codified in customary laws for seed governance. Western science, by contrast, has forgotten this element of "feeling", which has allowed people to go against nature.

The Quechua concept of "Uyway" or "crianza", which can be understood as mutual caring (Allen, 2017) or reciprocal nurturing applies strongly to seeds, but is broader than seeds (also applies to animals, land, rocks, etc.). The term means both nurturing seeds and being nurtured or

Fig. 4.1 Quechua farmers in the Potato Park Peru celebrate the spirit of the potato. Asociación ANDES

> nourished by seeds, in a circular way, encompassing the concept of reciprocity or *ayni*. These are normative principles that people continue to use. Seeds are viewed as children: nourished, protected and treated as part of the family, who will help the family when they grow; and when plants are fully grown they are respected as elders.

In the Potato Park, native seeds are mainly accessed through seed saving, exchange with other farmers or local barter (ANDES, 2016). The sustainability and adaptability of farming and food system depends on farmers having the right to freely exchange seeds, develop new varieties and maintain their rights over traditional varieties (Sayre et al., 2017). When potatoes are harvested, tubers are first selected for use as seed (Sayre et al., 2017). Women play a key role in selection, conservation, storage and management of native seeds, using techniques learned from their mothers; they participate in local seed fairs and have extensive knowledge about the uses and properties of native varieties (ANDES, 2016).

Andean farmers also make significant use of biocultural indicators such as observations of stars, wild plants and animals, interpreted to determine precise dates for planting, harvesting and rituals (Sayre et al., 2017). As Potato Park seed expert Lino Mamani has explained,

> *we have indicators from wildlife so we can do forecasting to know when we can plant; we do seed selection when wildlife such as a fox, or the sky, tells us – this is very important for seed selection. We have experts that read the signs and farmers have their own knowledge. When we do planting neighbours help each other, so it is collective.* (Swiderska & Ryan, 2021).

When potatoes are harvested, some tubers are put in the Potato Park's community seed bank, where seeds are stored "using a type of selection which reflects our taxonomy" (Potato Park expert, in Swiderska & Ryan, 2021). The seed bank enables farmers to access seeds when they lose a particular variety (ANDES, 2016). The seed diversity is also conserved for the future and for sharing with other communities (Swiderska & INMIP, 2017). The Potato Park is planning to establish a new community seed enterprise to support its seed conservation work; this will focus on botanical (i.e. in vitro) seed that is disease-free and can be stored in the seed bank for 50 years (Swiderska & INMIP, 2017). Guided by the Andean principle of reciprocity, the enterprise will focus on multiplying seeds for distribution to communities, rather than selling large quantities for profit.

CWRs and other semi-wild and wild plants are widely distributed throughout the Potato Park and grow naturally among cultivated varieties. The farmers know that wild potatoes growing near culti-vated fields improve the resilience and yield of cultivated varieties (Swiderska & INMIP, 2017). They know the importance of "combining them with domesticated potatoes so they can converse", (Lino Mamani, in Swiderska & Ryan, 2021). As Nasario Quispe, another Potato Park expert, explained, wild potatoes "produce their own tribe – some produce potatoes good for drying, others for other uses. Keeping all this diversity ensures we have enough food" (Swiderska & Ryan, 2021).

Elderly Quechua women, who take care of livestock grazing, plant potatoes in animal corrals where the seeds of wild potatoes germinate in animal dung, so that CWRs can fortify their domesticated varieties through cross-breeding (Swiderska & INMIP, 2017). CWRs are also used for rituals and food in times of famine, and to make freeze-dried potatoes for long-term storage. The Potato Park also has wild populations of *oca* tubers (Oxalis tuberosa), *tarwai* (lupins), *mashua* (similar to turnip) and passion fruit (Swiderska & INMIP, 2017).

Box 4.4: The potato park as genetic reserve for in-situ conservation
The Potato Park genetic reserve for in-situ conservation of CWRs was launched by the international non-profit Crop Trust and the government of Peru in 2017. CWRs are vital for use as fuel, medicine, food and fodder, and as a source of climate-adaptive traits, as they can respond much faster to climate change than cultivated varieties and have higher genetic diversity.

The Potato Park genetic reserve covers the whole landscape, especially the areas of higher elevation. Its main objective is to maintain genetic diversity in wild populations so they can keep evolving. Wild relatives are regarded as the grandparents of domesticated species, but there is still "communication" between wild and domesticated varieties. Two of the wild potato species known to occur in the Park are very closely related to domesticated species, and can be crossed with domesticated potatoes to produce offspring that have improved tolerance to pests, frost, high temperatures and drought.

To create the genetic reserve, the Potato Park collected baseline data by mapping CWRs and hotspots of genetic diversity at high altitude in all five communities, with support from Asociación ANDES. Farmers collected information on the range and population density of CWRs, and how frequently cultivated potatoes are planted near wild populations. Laboratory analysis has also been done, for instance on germination potential, which helps in understanding pollination and whether CWR populations are viable sizes. Three wild potato species in the Potato Park—*Solanum bukasovii*, *S. acaule* and *S. raphanifolium*—grow naturally in a range of environments. Populations of *S. acaule* are found in high densities where livestock graze, indicating dispersal by animals.

Ancestral knowledge of farmers about CWRs has also been collected, and used to develop management plans for the genetic reserve alongside Western science. Rules for conservation and management of wild relatives already exist in the Quechua normative principles of the Potato Park communities, and in many Indigenous communities, and so do not need to be re-invented by conservationists.

Source Presentation by Eve Allen, INMIP Exchange 2017 (Swiderska & INMIP, 2017)

4.4   The Ayllu System of Andean Seed Governance

The governance of the Potato Park landscape and seed system is based on the Andean pre-colonial concept of *sumaq causay*, or the holistic wellbeing of both people and nature (Sayre et al., 2017; Swiderska et al., 2020). In the Quechua and Aymara holistic worldview, the world is made up of three communities or Ayllus (Aniseto Ccoyo, Saccaca Community, in Swiderska & INMIP,2017; Marisol de la Cadena, 2015):

- Runa Ayllu: humanity and all domesticated elements (plants, animals, water)
- Sallka Ayllu: wild animals and plants, and all elements outside human control
- Auki Ayllu: the sacred and the ancestors.

These three Ayllus must be in balance to achieve holistic wellbeing (see Fig. 4.1). Balance is achieved through reciprocity between the Ayllus, which "glues the three communities together", notes Aniseto Ccoyo of the park's Saccaca Community (in Swiderska & INMIP, 2017). Ccoyo went on to say: "Good living only happens when there is harmony between the three Ayllus. The most sacred element of all this is the Pachamama (Mother Earth). This way of organising the three Ayllus is very old. It is very much alive in many communities, but they do not use it to organise their space as we are doing here" (in Swiderska & INMIP, 2017). Many native scholars define Ayllu as an "enlarged community" encompassing all the living and non-living organisms located in a given territory.[2]

The Ayllu system means that seed governance in the Potato Park is not based on human exceptionality—humans are not the only or the highest authority. As described above, other elements—the wild and the sacred (such as the sacred mountains)—also have souls and identities, and play an institutional role in the governance of seeds and crops. All elements, not just humans and food-producing habitats, contribute to creating rules for seed governance and conservation. Humans are part of a larger community and have to negotiate with the other members, the sacred and the wild. Thus, governance is not a human-centric approach, but is also guided by nature and spirituality.

This holistic worldview is similar in other Indigenous cultures, for instance in Asia and Africa. The common element is that people are not separated from nature—this is what distinguishes a biocultural heritage approach to seed governance. Seeds are part of biocultural heritage, embodying the indivisibility of nature and culture. But in other communities that are already immersed in the Western worldview, only humans play a role in governance, and this allows economic goals to dominate. IPRs that allow private ownership of seeds arise from the separation of humans and nature, and run counter to Indigenous peoples' holistic governance which includes non-humans in the larger construct of society (Fig. 4.2).

[2] https://incayllu.blogspot.com/p/blog.html.

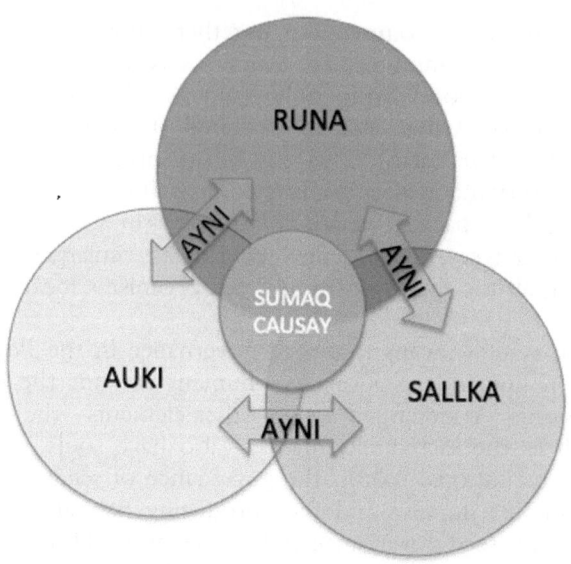

Fig. 4.2 Holistic wellbeing (*sumaq causay*) as balance between the three communities, or Ayllus, of the Andean worldview

4.5   ENRICHING SEED HERITAGE: POTATO REPATRIATION AND THE INTER-COMMUNITY AGREEMENT

In 2004, the Potato Park Association, ANDES and the International Potato Centre in Lima (CIP) signed the Agreement for the Repatriation, Restoration and Monitoring of Agrobiodiversity of Native Potato and Associated Community Knowledge Systems. Under this historic five-year agreement, the CIP gene bank returned 410 germ-free native potato cultivars to the Potato Park communities, for food security and in-situ conservation. CIP scientists had collected the varieties from Potato Park communities in the 1960s; subsequently the cultivars had disappeared from the communities through genetic erosion.

This repatriation agreement, probably the first from a gene bank to communities, recognised the importance of in-situ–ex-situ linkages for food security and climate adaptation (Stenner et al., 2016). A second five-year agreement, signed in 2010, involved collaborative research activities to monitor and test the repatriated potato varieties, linking traditional knowledge and science.

The agreement helped restore seed diversity as well as associated Indigenous knowledge and beliefs embedded in the seeds. It enhanced food security, resilience to climate change and livelihoods. It also contributed to the protection of Potato Park communities' rights over their native potato varieties and ancestral knowledge—a key objective is ensuring that genetic resources and knowledge remain under the custody of the communities and do not become subject to IPRs. Through this agreement, communities' understanding of their rights has grown, along with their capacity to protect those rights through community knowledge registers, developed with support from ANDES. The agreement also led to the development of the Potato Park's inter-community agreement for benefit sharing (Stenner et al., 2016).

The Inter-Community Agreement (see Box 4.2) provides the foundation for collective decision-making, good governance and social cohesion among the five communities of the Potato Park. It recognises collective custodianship over Indigenous knowledge and seeds. Everyone has the right to freely access knowledge (barring sacred knowledge) and seeds, and the obligation to maintain their free flow among Potato Park and neighbouring communities, as well as to transmit them to future generations (ANDES et al., 2011).

The Potato Guardians collective (see Box 4.2) was established to manage potato repatriation. The in vitro seeds are treated in a small lab before being multiplied in a greenhouse, where the whole Potato Park collection is stored. The seedlings then go to a net-house and are distributed to the communities, which can request particular varieties. Farmers manage the potatoes in their fields, and each community has their own potato collection (Representative of Potato Guardians Collective, in Swiderska & INMIP, 2017).

4.6   Conclusions and Recommendations

Indigenous peoples have domesticated, conserved and improved much of the world's genetic diversity, adapting seeds to changing environmental conditions through their knowledge and cultural values. In this way, they have ensured food security and biodiversity conservation for millennia. Their seed systems are at the forefront of genetic evolution and adaptation to climate change, sustaining CWRs that are the first to develop resilient properties, and actively harnessing these properties to enrich domesticated crops.

Safeguarding this genetic diversity is crucial for global food security, particularly given the increasing homogenisation of agri-food systems and growing climatic extremes and unpredictability. Genetic diversity is also crucial for Indigenous peoples' own food security, to reduce the risk of crop failure, and ensure stable production, dietary diversity and nutrition. Perhaps even more importantly, it is a key source of cultural identity, pride, social cohesion and spiritual wellbeing.

Across the world, so-called "informal" Indigenous seed systems are in fact actively governed through customary laws, principles and beliefs. Key common principles across continents include sharing and reciprocity (the obligation to share and exchange seeds), balance with nature, and solidarity (the obligation to help those in need). The sacredness of seeds as givers of life is also enshrined in many Indigenous cultures. But the role of Indigenous principles, values and cosmovision in sustaining genetic diversity is largely overlooked in, and often undermined by, formal seed policies (Graddy, 2013; Wattnem, 2016). Analysis of farmers' and formal seed systems shows important complementarity between the two, but very few countries reflect that in their seed policies (Almekinders & Louwaars, 2002).

The integrity of seeds as biocultural heritage is best protected through a biocultural rights-based approach to seed governance. Biocultural rights are a community's long-established right, in accordance with its customary laws, to steward its lands, waters and resources. They are not simply claims to property, but the collective rights of communities to carry out traditional stewardship roles, as conceived of by Indigenous ontologies (Bavikatte & Bennett, 2015). An approach based on these rights goes beyond the concepts of food or seed sovereignty by explicitly recognising the cultural and spiritual nature of Indigenous seed governance.

Increasingly, biocultural rights are being recognised in international environmental law (Bavikatte & Bennett, 2015). Recognising these rights in national and international law means recognising concepts like biocultural heritage and protecting customary legal systems. In Indigenous cultures, biocultural rights include the rights of Mother Earth, rivers or lakes—which have been formally recognised in a few cases—and the rights of seeds.

Protecting Indigenous seed systems and implementing a biocultural rights-based approach demands:

- reforming national seed policies to remove the threats to Indigenous seed systems (IPR and non-IPR related) and promote mutually supportive formal and informal seed systems, and reforming other mainstream policies that threaten Indigenous seed systems and biocultural heritage;
- actively engaging Indigenous peoples in defining policies and laws for seed governance, so that these recognise and reinforce existing customary laws that seek to preserve genetic diversity and ensure seed access;
- establishing a global network of Biocultural Heritage Territories for effective long-term protection of Indigenous seed systems in centres of origin and diversity of crops, linking wild and domesticated gene pools, through decolonising action-research processes, building on the successful Potato Park model;
- supporting the establishment of community seed banks based on Indigenous peoples' knowledge and rules for storing and accessing seeds, which are diverse and adapted to local cultures and ecologies (such as traditions for maize seed storage in Mexico and China, which predate scientific methods);
- supporting Indigenous plant breeding systems which have existed for millennia and are similar to evolutionary plant breeding, along with "farmer field schools" based on Indigenous cosmovision and ways of knowing, going beyond farmers' fields or particular technologies.

The Covid-19 pandemic has shown the critical importance of Indigenous seed and knowledge systems for resilience to shocks: they have proved far more resilient than those based on Western science and global markets. This points to an opportunity to re-evaluate seed governance and promote more pluralistic governance systems.

REFERENCES

African Biodiversity Network and Gaia Foundation. (2015). *Celebrating African rural women: Custodians of seed, food and traditional knowledge for climate change resilience.* https://www.gaiafoundation.org/app/uploads/2015/11/Celebrating-African-Rural-Women.pdf

AFSA & GRAIN. (2018). *The real seed producers: Small-scale farmers save, use, share and enhance the seed diversity of the crops that feed Africa.* Alliance for

Food Sovereignty in Africa (AFSA) and GRAIN. http://www.grain.org/e/6035

Allen, C. J. (2017). Connections and disconnections: A response to Marisol de la Cadena. *HAU: Journal of Ethnographic Theory, 7*(2). https://doi.org/10.14318/hau7.2.003

Almekinders C., & Louwaars, N. (2002). The importance of the farmers' seed systems in a functional national seed sector. *Journal of New Seeds, 4*(1–2). https://doi.org/10.1300/J153v04n01_02

ANDES (Peru), the Potato Park Communities & IIED. (2011). *Community biocultural protocols: building mechanisms for access and benefit sharing amongst the communities of the Potato Park based on customary Quechua norms*. Summary report. IIED, ANDES. Retrieved May 2021 from https://pubs.iied.org/g03168

Argumedo, A., & Pimbert, M. (2005). *Traditional resource rights and indigenous people in the Andes*. IIED, ANDES. Retrieved May 2021 from https://pubs.iied.org/sites/default/files/pdfs/migrate/14504IIED.pdf

Argumedo, A., Swiderska, K., Pimbert, M., Song, Y., & Pant, R. (2011). *Implementing farmers' rights under the FAO International Treaty on PGRFA: The need for a broad approach based on biocultural heritage*. IIED. Retrieved May 2021 from https://pubs.iied.org/sites/default/files/pdfs/migrate/G03077.pdf

Asociación ANDES. (2016). *Resilient farming systems in times of uncertainty: Biocultural innovations in the Potato Park, Peru*. Country Report. IIED. Retrieved May 2021 from https://pubs.iied.org/14663iied

Bavikatte, K., & Bennett, T. (2015). Community stewardship: The foundation of biocultural rights. *Journal of Human Rights and the Environment, 6*(1), 7–29.

Bellon, M. R. (1996). The dynamics of crop infraspecific diversity: A conceptual framework at farmer level. *Economic Botany, 50*(1), 26–39. https://doi.org/10.1007/BF02862110

Borelli, T., Hunter, D., Powell, B., Ulian, T., Mattana, E., Termote, C., ... Engels, J. (2020). Born to eat wild: An integrated conservation approach to secure wild food plants for food security and nutrition. *Plants, 9*(10), 1299. https://doi.org/10.3390/plants9101299

CIP. (2008). *The world potato atlas*. International Potato Centre. Retrieved May 2021 from https://research.cip.cgiar.org/confluence/display/wpa/Home

De la Cadena, M. (2015). *Earth beings: Ecologies of practice across Andean worlds*. Duke University Press.

FAO. (2018). *High-level expert seminar on indigenous food systems: Building on traditional knowledge to achieve zero hunger*. Retrieved May 2021 from http://www.fao.org/indigenous-peoples/food-systems/en/

Gill, T. B., Bates, R., Bicksler, A., Burnette, R., Ricciardi, V., & Yoder, L. (2013). Strengthening informal seed systems to enhance food security in Southeast Asia. *Journal of Agriculture, Food Systems and Community Development, 3*(3), 139–153. http://dx.doi.org/10.5304/jafscd.2013.033.005

Graddy, T. G. (2013). Regarding biocultural heritage: In situ political ecology of agricultural biodiversity in the Peruvian Andes. *Agriculture and Human Values, 30,* 587–604.

Gutierrez, L. M. (2016). *The political ontology of seeds: Seed sovereignty struggles in an Indigenous resguardo in Colombia.* Ph.D. dissertation, University of North Carolina at Chapel Hill Graduate School. https://cdr.lib.unc.edu/concern/dissertations/ks65hd78f

Hunter, D., & Heywood, V. (Eds.) (2011). *Crop wild relatives. A manual of in situ conservation.* Earthscan.

Hunter, D., Borelli, T., Beltrame, D. M., Oliveira, C. N., Coradin, L., Wasike,V, ... Madhujith, T. (2019). The potential of neglected and underutilized species for improving diets and nutrition. *Planta, 250*(3), 709–729.

Jiggins, J. (2017). Gender and agricultural biodiversity. In D. Hunter, L. Guarino, C. Spillane, & P. McKeown (Eds.), *Handbook of agricultural biodiversity* (pp. 525–534). Routledge.

Maxted, N., Hunter, D., & Ortiz, R. (2020). *Plant genetic conservation.* Cambridge University Press.

Munyi, P., & de Jonge, B. (2015). Farmers' and breeders' rights: Bridging access to, and IP protection of, plant varieties in Africa. *The African Journal of Information and Communication,* 16.

Pant, R. (2012). *Heritage on the edge: protecting traditional knowledge and genetic resources in the Eastern Himalayas, India.* IIED. Retrieved May 2021 from https://pubs.iied.org/g03442

Pilgrim, S., & Pretty, J. (2010). *Nature and culture: Rebuilding lost connections.* Earthscan.

Posey, D. et al. (1996). *Traditional resource rights: International instruments for protection and compensation for Indigenous peoples and local communities.* IUCN.

Rowland, D., Ickowitz, A., Powell, B., Nasi, R., & Sunderland, T. (2017). Forest foods and healthy diets: Quantifying the contributions. *Environmental Conservation, 44,* 102–114. https://doi.org/10.1017/S0376892916000151

Samuel, B., & David, N. (2007). Approaches to food security and agriculture: options for constructing and agenda that builds on cultural diversity. In B. Haverkort & S. Rist (Eds.), *Endogenous development and bio-cultural diversity: The interplay of worldviews, globalization and locality.* COMPAS.

Sayre, M., Stenner, T., & Argumedo, A. (2017). You can't grow potatoes in the sky: Building resilience in the face of climate change in the Potato Park of

Cuzco, Peru. *Culture, Agriculture, Food and Environment, 39*(2), 100–108. https://doi.org/10.1111/cuag.12100

Sievers-Glotzbach, S., Euler, J., Frison, C., Gmeiner, N., Kliem, L., Maze, A., & Tchersich, J. (2021). Beyond the material: Knowledge aspects in seed communing. *Agriculture and Human Values, 38,* 509–524. https://doi.org/10.1007/s10460-020-10167-w

Stenner, T., Argumedo, A., Ellis, D., & Swiderska, K. (2016). Potato Park-International Potato Center-ANDES Agreement: Climate Change Social Learning (CCSL) case study on the repatriation of native potatoes. In M. Van Epp & B. Garside (Eds.), *Solving wicked problems: Can social learning catalyse adaptive responses to climate change? A compendium of case studies.* IIED Retrieved May 2021 from https://pubs.iied.org/sites/default/files/pdfs/migrate/17398IIED.pdf

Swiderska, K., Argumedo, A., & Pimbert, M. (2020). *Biocultural heritage territories: Key to halting biodiversity loss.* IIED Briefing. IIED. May 2021 from https://pubs.iied.org/17760iied

Swiderska, K., Argumedo, A., Song, Y., Rastogi, A., Gurung, N., & Wekesa, C. (2018). *Biocultural innovation: The key to global food security?* IIED Briefing. IIED Retrieved May 2021 from https://pubs.iied.org/17465iied

Swiderska, K., Argumedo, A., Song, Y., Li, J., Pant, R., Herrera, H. … Vedavathy S (2009). *Protecting community rights over traditional knowledge: Implications of customary laws and practices: Key findings and recommendations (2005–2009).* IIED. Retrieved May 2021 from https://pubs.iied.org/14591iied

Swiderska, K., & INMIP. (2017). *Resilient biocultural heritage landscapes for sustainable mountain development.* International Network of Mountain Indigenous People (INMIP). https://pubs.iied.org/14670iied

Swiderska, K., & Ryan, P. (2021). *Indigenous food systems, biocultural heritage and the SDGs: challenges, interdisciplinary research gaps and empowering methodologies.* IIED. Retrieved May 2021 from https://pubs.iied.org/20191IIED

Swiderska, K., Song, Y., Li, J., Reid, H., & Mutta, D. (2011). *Adapting agriculture with traditional knowledge.* IIED Briefing. IIED. Retrieved May 2021 from https://pubs.iied.org/17111iied

Swiderska, K., & Stenner, T. (2020). *The Maize Park Biocultural Heritage Territory in Lares, Peru: Case study guidance on biocultural heritage territories.* International Network of Mountain Indigenous Peoples (INMIP). Retrieved May 2021 from https://pubs.iied.org/g04447

Walshe, R., & Argumedo, A. (2016). Ayni, Ayllu, Yanantin and Chanincha: The cultural values enabling adaptation to climate change in communities of the Potato Park, Peruvian Andes. *GAIA, 25*(3), 166–173. https://doi.org/10.14512/gaia.25.3.7

Wattnem, T. (2016). Seed laws, certification and standardisation: Outlawing informal seed systems in the Global South. *The Journal of Peasant Studies, 43*(4), 850–867. https://doi.org/10.1080/03066150.2015.1130702

Wilson K. B. (2009). *Foreward, in 'Where our Food Comes From: Retracing Nikolay Vavilov's quest to end famine' by Gary Paul Nabhan*. Island Press /Shearwater Books.

CHAPTER 5

The Diversity of Seed-Saving Governance and Sharing Systems in contemporary Japan

Ayako Kawai

Abstract Crop diversity in Japan is on the ebb, eroded by factors such as the rise of industrialised agriculture, a shrinking and ageing population of farmers, and a dearth of knowledge transmission between generations. However, thousands of Japanese farmers follow a practice vital to fostering agrobiodiversity: seed saving. Using a qualitative case study approach, Ayako Kawai tracked diverse seed governance and sharing systems across four groups of producers: traditional, organic and 'lifestyle' farmers and local community members. She found differences in the ways seeds are valued—cultural, economic, rights-based, familial or personal—that influence approaches to saving and sharing seeds. Organic and traditional farmers and community growers, for instance, tightly regulate seed distribution, while part-time producers are far keener to actively share seeds. That could, notes Kawai, create a dilemma if broader access to genetic resources becomes a general priority. Yet she concludes that a plurality of practices, like crop diversity itself, builds in resilience by spreading risk and offering a range of responses to future uncertainties.

A. Kawai (✉)
Research Institute for Humanity and Nature, Kyoto, Japan

© The Author(s) 2022
Y. Nishikawa and M. Pimbert (eds.), *Seeds for Diversity and Inclusion*,
https://doi.org/10.1007/978-3-030-89405-4_5

Keywords Intellectual property rights · Seed saving governance ·
Seed sharing · Trust · Values

5.1   INTRODUCTION

The availability of seeds, and the way farmers access them, are key to
the success of agrobiodiversity conservation and management (Hodgkin
et al., 2007; Jarvis et al., 2011). And informal seed systems are important
in this context: they are generally recognised as helping to maintain agrobiodiversity as well as providing farmers with better access to seeds (FAO,
2010). Yet the socio-cultural factors involved in seed distribution under
such systems, such as social ties, social status and cultural norms (Badstue
et al., 2006; Jarvis et al., 2011), have to date been largely ignored by
researchers.

In Japan, the rise of agricultural industrialisation is damaging crop
diversity. Since the 1960s, government policy reform has facilitated mono-
cultural mass vegetable production and replaced local varieties with
commercial hybrids. The limited market value of local varieties and the
fall in numbers of farmers have further contributed to the loss of crop
diversity (FAO, 2010). As in Italy (Negri, 2003) and other developed
countries (Pautasso et al., 2013), the farmers who save local varieties
in Japan constitute an ageing population; the limited transmission of
their knowledge to younger generations poses another challenge for the
conservation of local varieties. While public interest in agrobiodiversity is
rising gradually in response to recent changes in seed-related laws, little is
known about seed saving in Japan—neither the groups involved, nor the
values that motivate them.

This chapter aims to redress the balance by exploring the values and
motivations of Japanese seed savers, with particular focus on seed sharing.
I compare four types of seed savers:

- traditional farmers (those who have received seeds and learned seed
  saving from their family members)
- non-traditional local community members
- full-time organic farmers
- lifestyle farmers (those who farm as a lifestyle choice and do not
  derive the majority of their income from farming).

Since no survey has been conducted in Japan to gauge the number of seed savers, it is difficult to provide a breakdown of the different groups engaging in the practice across the country. According to data reported by the Japan Organic Agriculture Association (JOAA, 2010) and Mokichi Okada Association Nature Farming and Culture Foundation (MOA, 2011),[1] however, some 7200 organic farmers in Japan save seeds.

This research is based on a qualitative case study approach, involving 62 semi-structured interviews and 44 days of participant observations in 14 prefectures across Japan through 2016–2018. My findings are as follows.

5.2 TRADITIONAL FARMERS

I found that traditional farmers save heirloom seeds, learning the skills from family members. They regarded seeds received from parents or in-laws as an ancestral heritage, motivating them to continue in the practice. For instance, some traditional farmers in Akka hamlet, Iwate prefecture, were committed to saving seeds because they saw them as such an inheritance from their ancestors. They felt that they should not *tayasu* (relinquish) those seeds: that would be an abdication of responsibility, and would mean the end of that variety.

Hiroaki Egashira, an academic and the president of a local crop-diversity conservation group, also noted during an interview that some traditional farmers continue to save seed in Yamagata prefecture because they "feel sorry for ancestors, losing what they have transmitted from generation to generation". Similarly, Kazuya Takahashi, a merchant specialising in traditional vegetables, reported that many traditional farmers felt deeply responsible for maintaining inherited seeds. Some, Takahashi noted, said that the person who first saved the seed "entrusted the seeds" to them, and that the seeds reminded them of the ancestral seed saver's face.

Traditional farmers often restricted their distribution of heirloom seeds. The most extreme cases were farmers who saw seeds as *mongai-fushutsu*

---

[1] The MOA (2011) estimated the number of full-time organic farming households across Japan as 12,000. The JOAA (2010) reported that 60% out of 155 interviewed organic farmers save vegetable seeds. I estimated the number of seed-saving full-time organic farmers by multiplying 12,000 by 0.6.

(meaning never given to others), except in regard to family members. For example, Kurofuji-kyūri, a variety of cucumber, has been maintained by an individual in Yamagata prefecture who has been faithful to the precept that it be conserved only by family members (Egashira, personal communication). Even without such explicit family rules, traditional farmers tend to feel hesitant about sharing their seeds. Takahashi observed that it was often difficult for non-traditional local savers to acquire seeds while the traditional farmers saving them were still living. Even within the local community, seed sharing was found to be patchy.

Reports of disrespectful acts by people receiving seeds also made traditional farmers cautious about extending their sharing more widely. In Iwate prefecture, for example, one person who received seeds of a local turnip variety registered a trademark for it using its recognised regional name. That act effectively prevented traditional farmers from selling their turnips under the name, resulting in a loss of potential income.

This study's finding that traditional farmers valued seed saving largely because of the ancestral connection echoes others in South Africa (van Niekerk & Wynberg, 2017) and Peru (Tobin et al., 2018). But it also reflects a narrative of ancestor worship and related obligations that is deeply rooted in Japanese culture, especially within the traditional family system. The challenge for traditional farmers to find a successor may not only be an issue to do with human resources. It also points to the fact that younger generations do not place the same significance on the narrative of ancestoral connection, leading to diminished motivation for saving heirloom seeds. Because of this trend, the transmission of seeds and knowledge from traditional farmers to the younger generation is an even more pressing issue.

Yet some traditional farmers have found ways of surmounting such barriers. For example, in Yamagata prefecture some traditional farmers who were ageing, and had no family members willing to continue cultivation, started to openly share their *mongai-fushutsu* seeds with local community members (Egashira, personal communication). As familial transmission ebbs, the significance of local community members' interest in conserving local varieties has increased.

5.3 Non-Traditional Local Community Members

In this group—non-traditional local community members saving local varieties of seeds transmitted outside their families—those I interviewed tended to receive the seeds from traditional farmers or, in some cases, gene banks or seed companies. Individuals in this group valued local varieties because it strengthened their identity by effectively connecting them to local cultural history.

Some I interviewed valued local variety seeds because of their part in conserving local culture and traditional landscapes. Akari, head of the local variety conservation group in Akka hamlet, felt that the surrounding landscape would be irrevocably altered if farmers stopped cultivating local varieties, and started seed saving on her own. Similarly, locals in Aomori prefecture revived the cultivation of a local spring onion variety, feeling that it represents local identity. Their stated goal was to "ensure that children in one hundred years can proudly say 'This is our spring onion'". One interviewee regarded such varieties as teaching materials, allowing young people to recognise local assets and meet those keen to protect them. He also noted that such learning fostered students' attachment to their hometown. Local varieties are thus strongly tied to regionalism and local identity.

Within the non-traditional local community, people often limited seed distribution to non-community members for a range of reasons: to maintain a price premium, protect the unique character of the region, or ensure high-quality production. In Aomori prefecture, a small number of contracted farmers handled seed distribution as a way of maintaining a premium price. They found it challenging to make a living from producing local varieties, and so generated higher profits by restricting the number of growers (and so the overall size of harvests), which rendered the variety exclusive. In Iwate prefecture, Akari designated seeds of the radish Akka daikon *mongai-fushutsu*, restricting seed sharing to residents of the hamlet. Very few of the resident farmers still cultivate it, which led Akari to fear that growers elsewhere might take up the variety—and so erode its unique connection to Akka. Interestingly, traditional farmers I spoke to, however, did not restrict seed sharing to this degree.

In Tokyo, one founder of a local variety conservation group restricted the distribution of local seed varieties and seedlings to locals. He thought it irresponsible to casually give away seeds because it was important to harvest "the original fruits". He commissioned local professional farmers to grow and reproduce the varieties distributed. In general, people I interviewed within this group valued the authenticity and uniqueness of a variety, so they restricted access to them to protect those values.

The people I interviewed in this overall group often created their own rules to limit seed distribution, and sometimes asked traditional farmers to comply with them. Such rules allowed locals to protect their intellectual property rights, which are different from formal regulations that mainly protect professional plant breeders. For instance, an organic farmer I interviewed, from Morioka city in Iwate prefecture, was interested in saving seeds of the radish on from Akka, which is in his prefecture; but Akka's local conservation group leader claimed the radish seeds to be *mongai-fushutsu*, so the farmer could not ask for them. In this sense, informally agreed rules enabled locals to effectively protect their intellectual property rights by refusing outsiders access to the seeds.

5.4   Organic Farmers

Organic farmers are full-time working farmers who rely mainly on seed saving for their crops. Those I interviewed were highly interested in saving local variety seeds, but often struggled to find or access them. In some cases, local varieties have disappeared from their areas. They also reported that identifying and building trust with traditional seed savers was challenging. Given these difficulties, organic farmers sought local varieties from neighbouring regions or purchased non-hybrid seeds from seed companies.

Through saving seeds, organic farmers aimed to develop their own varieties adapted to local environmental conditions. One young farmer who sourced seeds of local varieties from local companies mentioned that once a variety developed new traits under cultivation, he would rename it using his own surname. Another, older organic farmer, who bred a variety highly praised by the owner of a seed company, was not willing to register it for intellectual property rights. He suggested that this was a norm among his fellow organic farmers.

The organic farmers I interviewed generally felt reluctant to distribute their seeds widely, fearing careless treatment. They preferred to entrust (*takusu*) seeds to those they had close relationships with. A person so trusted was responsible for taking good care of the seed and not distributing it widely either—to protect farmers' rights as well as their emotional attachment to their self-bred varieties. One farmer indicated that he was very concerned that the crops he gave to a trusted farmer "lived well". Like some other organic farmers, he effectively viewed the seeds as his children, and felt that it was his duty to ensure their wellbeing.

The seed-sharing strategy of organic farmers thus differed from those of traditional farmers, and of non-traditional community members. Still, their caution about distributing seeds functioned as a kind of informal protection of intellectual property rights. Organic and lifestyle farmers did not necessarily ascribe the same values as to seed sharing, for instance, and this created friction between them. Several organic farmers I interviewed mentioned that they felt disappointed when they shared seeds with casual savers who were not committed to the practice. A case showing a similar preference to give seeds to those enthusiastic about multiplying seeds was reported in South Africa (van Niekerk & Wynberg, 2017).

5.5   LIFESTYLE FARMERS

Lifestyle farmers engage in farming as a lifestyle choice, and do not earn major income from farming. Those I interviewed saw seed saving as a hobby and most did not show any special interest in saving the seeds of local varieties. They mainly bought seeds from seed companies or acquired them from informal exchange networks.

Lifestyle farmers that I interviewed followed natural farming methods[2] that are reliant on ecological processes and minimise human interventions. Under this approach, they leave plants on farm after they wither, allowing the seeds to fall in situ; natural germination is highly valued. Since these farmers did not, strictly speaking, conduct plant selection or regard their activity as plant breeding, they were not concerned about wide seed distribution.

The interviewees casually shared seeds with friends and similar-minded people—a practice very different from the restrictive strategies of the other three farming groups studied. Unlike organic farmers, none of the lifestyle farmers mentioned a sense of responsibility to maintain seeds that they received from others; nor did they worry about the state of seeds given to others. This was partly because lifestyle farmers' full-time jobs prevented them from fully committing to farming. This group also produced more seeds than they could use, and preferred to give them to

---

[2] Based on methods of the Japanese farmer-philosopher Fukuoka Masanobu, which were further developed by the farmer and author Yoshikazu Kawaguchi.

others than to dispose of them—believing, with the other groups, that each seed is a living entity and must be treated with respect.

Lifestyle farmers' active seed sharing inspired many recipients to take up the practice. Yet in this context, these farmers' values and methods could also spark conflict. For example, one organic farmer mentioned it as problematic when seed receivers at seed-sharing events multiplied those seeds and subsequently shared them at other events. He noted that organic farmers donate their seeds, which are not registered, out of goodwill and in the expectation that receivers will not widely distribute them.

5.6   Discussion and Conclusion

Sustainable seed management demands engagement by multiple actors. This study shows that seed saving in Japan is indeed carried out by multiple actors, each dealing with different types of seeds and conforming to different norms related to sharing them.

Figure 5.1 shows the four groups of Japanese seed savers introduced in this chapter, arranged according to two attributes: whether the act of sharing was open or restrictive, and whether the varieties maintained were integrated into local food culture, practices and traditions.

Fig. 5.1 Japanese seed savers: relative approaches to seed sharing, links to local culture

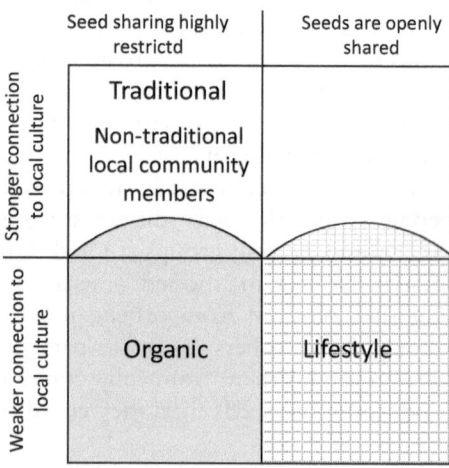

Lifestyle farmers were found to be keen to actively share seeds. While they concentrated primarily on non-local variety seeds, they also saved those of local varieties. By contrast, traditional farmers, non-traditional local community members and organic farmers followed norms and rules aimed at regulating seed distribution. Traditional and non-traditional local community members saved local varieties, while organic farmers saved both local and non-local. There were cases when farmers in the restrictive seed-sharing sphere occasionally widely shared their seeds, but these were exceptional and hence not reflected in the figure. (Open seed distribution of local variety seeds were sometimes carried out by local small seed companies, discussed in Chapter 10.)

The coexistence of diverse seed-saving practices with different rules and values reveals the plurality of seed governance. That diversity could in turn contribute to the conservation of crop diversity and food systems resilience in two ways.

The first is functional redundancy: by having multiple seed-saving practices, the loss of one could be compensated by others. While traditional farmers' practice in this regard was less likely to be sustained in the future compared with that of others, non-traditional community members and organic farmers were keen to save traditional farmers' seeds. The challenge lies in the restrictive seed-sharing attitudes and behaviours observed in traditional farmers, which could impede the effect of functional redundancy.

The second benefit of having diverse seed-saving practices is that in the face of future uncertainties, different groups of farmers may react in different ways. That range of responses could become a source of resilience. The four groups studied were on different economic footings (some saved seeds as part of their livelihoods, some not) and worked within different institutional settings (some were registered as farmers, some engaged in seed saving as an individual or as part of an organisation). Thus, different farmers may respond to the risks and challenges thrown up by ecological or social change in diverse ways. The farmers researched also had distinct motivations for saving seeds, and some were more committed to continue the practice than others. That may in turn influence the sustainability of each practice and the quality of genetic diversity maintained by each group of farmers in the context of future challenges.

My findings showed that trust is viewed as an important quality in seed receivers. Corresponding to the findings of Lone B. Badstue and colleagues (2007), and Jaci van Niekerk and Rachel Wynberg (2017), I found that seed transactions mainly took place between people with established social relationships. Some farmers and communities set stricter rules and memberships for seed transactions, often to protect savers' intellectual property rights regarding seeds. While having diverse seed exchange networks or social structures enhances the resilience of seed systems (Pautasso et al., 2013), this study showed that the plurality of seed governance may not necessarily ensure that farmers have better access to seeds.

The dynamic of different institutions and social networks influences the outcome of natural resource management (Bodin & Crona, 2009). My research reveals tension and conflict between different groups of seed savers, hinging on differing attitudes to seed sharing and a lack of communication about those values. Identifying and openly expressing implicit values regarding seed sharing among diverse actors may facilitate better seed transmission within and beyond such groups.

It is important to highlight that Japanese farmers' resistance to broad seed sharing is intertwined with a strong sense of commitment to seed saving. This, however, creates a dilemma when they pursue agroecological values that aim to broaden access to genetic resources. Understanding farmers' values around seed saving, and the underlying socio-cultural context, is critically important in developing crop-diversity conservation measures that are effective, inclusive and resilient.

Acknowledgements The research has been supported partially by Research Institute for Humanity and Nature (RIHN: A constituent member of NIHU), FEAST Project (No. 14200116).

References

Badstue, L. B., Bellon, M. R., Berthaud, J., Juárez, X., Rosas, I. M., Solano, A. M., & Ramírez, A. (2006). Examining the role of collective action in an informal seed system: A case study from the Central Valleys of Oaxaca, Mexico. *Human Ecology, 34,* 249–273. https://doi.org/10.1007/s10745-006-9016-2

Badstue, L. B., Bellon, M. R., Berthaud, J., Ramírez, A., Flores, D., & Juárez, X. (2007). The dynamics of farmers' maize seed supply practices in the Central Valleys of Oaxaca, Mexico. *World Development, 35,* 1579–1593. https://doi.org/10.1016/j.worlddev.2006.05.023

Bodin, O., & Crona, B. I. (2009). The role of social networks in natural resource governance: What relational patterns make a difference? *Global Environmental Change, 19*, 366–374. https://doi.org/10.1016/j.gloenvcha.2009.05.002

FAO. (2010). *The second report on the state of the world's plant genetic resources for food and agriculture*. FAO.

Hodgkin, T., Rana, R., Tuxill, J., Didier, B., Subedi, A., Mar, I., … Jarvis, D. (2007). Seed systems and crop genetic diversity in agroecosystems. In D. Jarvis, C. Padoch, & D. Cooper (Eds.), *Managing biodiversity in agricultural ecosystems* (pp. 77–116). Columbia University Press.

Jarvis, D. I., Hodgkin, T., Sthapit, B. R., Fadda, C., & Lopez-Noriega, I. (2011). An heuristic framework for identifying multiple ways of supporting the conservation and use of traditional crop varieties within the agricultural production system. *Critical Reviews in Plant Sciences, 30*, 125–176. https://doi.org/10.1080/07352689.2011.554358

JOAA. (2010). *Yūki nōgyō ni tsukau shubyō ni kansuru seisan, ryūtsū, riyō jittai chōsa hōkoku (2) – jika saisu wochūsin to shite* [Report on the status of production, distribution and utilization of seeds and seedlings in organic farming (2)—Focusing on seed-saving]. Retrieved August 5, 2021, from https://www.1971joaa.org/出版物-土と健康-書籍-dvd/調査事業報告書/#houkoku06

MOA. (2011). *Yūki nōgyō kiso dēta sakusei jigyō hōkokusho* [Report on the project of gathering data on organic farming]. Retrieved August 5, 2021, from https://moaagri.or.jp/manage/wp-content/themes/moaagri/pdf/hojojigyo/H22_yukikiso_houkokusho.pdf

Negri, V. (2003). Landraces in central Italy: Where and why they are conserved and perspectives for their on-farm conservation. *Genetic Resources and Crop Evolution, 50*, 871–885. https://doi.org/10.1023/A:1025933613279

Pautasso, M., Aistara, G., Barnaud, A., Caillon, S., Clouvel, P., Coomes, O. T., … Tramontini, S. (2013). Seed exchange networks for agrobiodiversity conservation: A review. *Agronomy for Sustainable Development, 33*, 151–175. https://doi.org/10.1007/s13593-012-0089-6

Tobin, D., Bates, R., Brennan, M., & Gill, T. (2018). Peru potato potential: Biodiversity conservation and value chain development. *Renewable Agriculture and Food Systems, 33*, 19–32. https://doi.org/10.1017/S174217051600284

van Niekerk, J., & Wynberg, R. (2017). Traditional seed and exchange systems cement social relations and provide a safety net: A case study from KwaZulu-Natal, South Africa. *Agroecology and Sustainable Food Systems, 38*(8), 861–889. https://doi.org/10.1080/21683565.2017.1359738

CHAPTER 6

Seed System Dynamics and Crop Diversity of Chinbaung in Myanmar

Mami Nagashima, Yoshiaki Nishikawa, Mya Shew, Ohm Mar Saw, Min San Tein, Makoto Kawase, Kazuo Watanabe, and Kenji Irie

Abstract In this field survey of seed system dynamics in Myanmar, the authors note that the country's dominant system of traditional agriculture faces pressure from the introduction of 'improved' varieties and shifts in policy. However, farmers—from small and subsistence growers to large-scale rice producers—continue to raise indigenous species. One is chinbaung, the collective term for several varieties in the genus *Hibiscus*. The authors traced differences in chinbaung cultivation and use among

M. Nagashima · K. Irie
Tokyo University of Agriculture, Setagaya, Tokyo, Japan

Y. Nishikawa (✉)
Ryukoku University, Kyoto, Japan
e-mail: nishikawa@econ.ryukoku.ac.jp

Mya Shew
Yezin Agricultural University, Naypyidaw, Myanmar

Ohm Mar Saw · Min San Tein
Ministry of Agriculture, Livestock and Irrigation, Naypyidaw, Myanmar

© The Author(s) 2022   91
Y. Nishikawa and M. Pimbert (eds.), *Seeds for Diversity and Inclusion*,
https://doi.org/10.1007/978-3-030-89405-4_6

places, and examined production systems in three villages in central arid zone, each sited in a geographically distinct locale. They found that a local festival popular with seed sellers has become a prime conduit for disseminating diverse genetic resources. Poe Yon, a guild of agricultural brokerage firms with hubs in cities across the country, meanwhile involves firms and farmers in a unique relationship that ensures broader distribution. Ultimately, the autonomy of farmers has enabled agrobiodiversity to thrive in Myanmar—a success, the authors note, that agricultural policymakers should heed.

Keywords Chinbaung · Festival · Hibiscus · Myanmar · Poe Yon · Vegetable

6.1   Introduction

Myanmar's crop diversity is rich, thanks to its subtropical location, varied climate and adherence to traditional agricultural practices. The country's policy of isolation from the international community under the junta regime (1962 to 2011, resuming in 2021) has meant that agriculture has been slow to modernise.

Recent development policies emphasising democratisation and economic growth have, however, sparked concern about the loss of genetic diversity (Thein et al., 2017; Tun & Than, 1996). Improved varieties have been actively introduced (Ministry of Agriculture and Irrigation, 2015) through international collaboration, and have also been imported commercially from countries including Thailand, China and Taiwan. It is increasingly common to find farmers growing just a few indigenous crop varieties in their backyards, not only because of the government's policy to promote improved varieties but also because many farmers wish to boost their income.

M. Kawase
Tokyo University of Agriculture, Atsugi, Japan

K. Watanabe
University of Tsukuba, Tsukuba, Japan

In light of this situation, this chapter focuses on the crops collectively called chinbaung in Myanmar. This Burmese term refers to several species in the genus *Hibiscus* which have long been grown across the country. We describe chinbaung's diversity in terms of biology, cultivation and culinary use, and analyse the processes through which that diversity is preserved, giving particular emphasis to different modes of seed procurement and distribution.

6.2   CHINBAUNG DIVERSITY: CULTIVATION AND USE

Chinbaung is cultivated on small and subsistence farms, and the species are also grown as a secondary crop by industrial farmers in Myanmar. For small farmers, the crop is a critical source of income as it requires less investment in fertilisers, pesticides and irrigation facilities. It is also easy to grow, and produces a stable yield for long periods, even under extremely dry conditions. By contrast, for large-scale rice farmers, chinbaung is a secondary source of income during the off-season. For agriculture in Myanmar generally, the crop underpins income stability, as other major crops are often at risk from flooding and drought.

The authors and colleagues collected 342 samples of chinbaung from around Yangon, Myanmar's largest city, as well as the central arid zone, the northeastern highlands and the southern delta from 2014 to 2018 (Domon et al., 2015; Nagashima, Yoshida et al., 2019). The findings for the five *Hibiscus* species are shown in Table 6.1.

H. sabdariffa is a *Hibiscus* species most commonly cultivated in Myanmar: it was found in all the administrative divisions surveyed. The second most frequently cultivated, *H. cannabinus*, was found in large numbers in the central arid zone. *H. radiatus* and *H. acetosella* were found more often in Chin State (a western mountainous state) than elsewhere. We surmised that the species grown in each region have been selected because of conditions such as temperature, precipitation and isolation due to the mountainous topology, which may have affected the establishment of different types (Mohamed et al., 2015; Sharma et al., 2016).

It was found that even when growing the same crop species, farmers sowed a mixture with different morphological characteristics, such as leaf size, stem height and rooting pattern. That implies that at least some of the plants can survive and flourish under varying, sometimes unexpected environmental conditions because they may have diverse adaptive

Table 6.1 Breakdown of collected samples of five *Hibiscus* species by district and state

Administrative district/division	H. sabdariffa	H. cannabinus	H. radiatus	H. acetosella	H. surattensis	Total
Kachin state	16	1	0	0	0	17
Sagaing region	37	8	2	1	1	49
Chin state	11	2	11	5	0	29
Shan state	39	3	2	2	0	46
Mandalay region	36	25	6	0	0	67
Magwe region	38	11	3	0	0	52
Kayah state	0	0	0	0	0	0
Kayin state	2	0	0	0	0	2
Bago region	15	6	3	0	1	25
Yangon region	14	2	0	0	0	16
Mon state	17	0	0	0	0	17
Ayeyarwardy region	11	0	3	0	0	14
Tanintharyi region	8	0	0	0	0	8
Total	244	58	30	8	2	342

traits, such as photoperiodic responsiveness, drought tolerance and/or resistance to cold or moisture, along with their range of morphological traits. The farmers told the authors that the practice reduces the risk of losing yields because of sudden deterioration in the weather.

We found that different chinbaung species were selected and used according to a range of purposes and taste preferences. *H. sabdariffa* is utilised as a food and fibre, and also has roles in medicine, dye, oil, pulp and fuel. Its versatility has made it a popular crop in resource-poor villages, and in small or garden farming. We discovered that *H. sabdariffa* exhibits five different morphological variations, while *H. cannabinus*, *H. radiatus* and *H. acetosella* are bimorphic (Nagashima et al., 2019). Furthermore, many minor differences have been found within each type.

6.3   How Farmers Use Chinbaung Seeds in Cultivation

Chinbaung is a frequent component of inter-cropping schemes, cultivated with rice in flood-prone areas in large plots up to 4 or 5 hectares (ha) in area, used in mixed cropping with other vegetables and pulses, or planted as a vegetable in domestic gardens. In the case of inter-cropping with rice, chinbaung serves as a subsidiary source of income. Chinbaung is also planted around oil crops such as sesame and sunflower to mitigate the risk of failure from drought, because the leaves as well as seeds can be harvested, which is not possible for the other oil crops.

H. sabdariffa blooms during the dry season around November, when there are under 13 hours of daylight, allowing the seeds to be harvested from December onwards. Many farmers don't purchase seeds, but collect them from their own fields to use for cultivation the following year, which reduces costs. This practice also maintains the selected crop traits, and can boost income via seed sales.

With such on-farm seed production, however, chinbaung must stay in the fields longer. The *H. sabdariffa* crop remains in the ground for more than a month after the leaves are harvested. Seed production is therefore limited to areas where farmers have plots large enough to accommodate that, on terrain less likely to be affected by events such as flooding. Sufficient income from the previous year's seed production is also key to the production of high-quality seeds, while continuous mutual agreement between seed growers and seed users will ensure the ongoing provision of such seeds. It is likely that the contractual production system contributes to the improvement of the quality of seeds and also the selection of varieties adapted to each area.

Similar contractual production systems are sometimes adopted by rice farming communities far from areas where chinbaung seed is produced. Lands around the township of Shwebo District in Sagaing Region, for instance, are sown with high-quality rice, with chinbaung intercropped as a source of additional income. Local farmers order the chinbaung seeds before sowing rice (around October to November) every year, purchasing them from seed suppliers, or seed producing farmers in the nearby village of Yao Te, who are paid when seed for the following year is ordered. This type of contract establishes a network in which specialised farmers produce high-quality chinbaung seed.

In some mountainous areas, by contrast, seeds are not collected after the vegetable harvest. For instance, in Taunggyi in Shan State, most farmers grow chinbaung for their own consumption, and allow seeds to fall and spontaneously germinate so they grow again the following year. Commercial suppliers in this area often sell seeds in small quantities, since the crops are largely confined to domestic gardens.

The circulation of *H. sabdariffa* seeds in Sagaing Region has revealed that seed farmers grow a mixture of varieties with different flower colours and morphological traits, as well as other chinbaung species. The farmers harvest the seeds without sorting or labelling them before shipping them to local seed suppliers; hence, any genetic mixture in seeds in this market can be human-induced.

In addition, some farmers around the township of Monywa District in Sagaing Region use different plots for the production of vegetables and chinbaung seed. These farmers produce seeds by sowing any remaining seeds in isolated plots after the vegetable growing season. The goal is not to select and maintain superior traits, but to secure income through the efficient use of vegetable plots.

6.4   Cultivation and Seed Procurement Methods in Relation to Flooding

Three adjacent villages (Ye Le Kyun, Yin Taw and Ywa Thit) in the central arid zone (Sagaing and Mandalay regions) were studied to explore residents' roles, relationships and attitudes vis-à-vis different vegetable and seed production systems. Chinbaung is widely cultivated across the regions. We also discovered related events such as the Shwe Kyun Pin Nat Festival (see Sect. 6.5), and bodies such as Poe Yon, a unique brokerage organisation that can influence these factors (see Sect. 6.6).

A comparison of the three villages revealed that geographical factors—their locations in relation to the river, each other and nearby cities—affected the forms of agriculture and methods of seed procurement used in each (see Fig. 6.1).

Ye Le Kyun (see Fig. 6.2) is sited on a towhead, or low-lying alluvial island, surrounded by river water during the rainy season. Farmers here produce a variety of seeds on a small scale and earn a profit by selling small quantities at high prices at Shwe Kyun Pin Nat Festival, held near the village; there are no large towns or seed suppliers nearby.

The fields of Yin Taw are not sowable in the rainy season. However, the village is close to two large cities, Sagaing and Mandalay, both of which have vegetable markets. Many farmers in Yin Taw focus on vegetable production, and have abandoned seed production altogether. They maximise the use of the land right up to the time when their fields begin to flood, and supply vegetables to the cities.

Fig. 6.1 Location of the three villages and Shwe Kyun Pin Nat Festival

Fig. 6.2 Low-lying Ye Le Kyun during the rainy season

Ywa Thit uses its flood-free location to its advantage, and engages in both dairy farming and seed production. Seeds are sold to neighbouring villages, seed suppliers and Poe Yon (see Sect. 6.6). The villagers also engage in the contractual production of high-quality seeds for villages that have difficulty in seed production because of the flood-swollen river.

Various forms of seed production at different scales and of cultivation methods were observed, contributing to the conservation of genetic diversity of chinbaung within and outside the area (Table 6.2).

6.5   RELATIONSHIP BETWEEN SEED DISTRIBUTION AND THE SHWE KYUN PIN NAT FESTIVAL

The Shwe Kyun Pin Nat Festival, held from 12 to 15 August (around the time of the full moon), is one of three major festivals held in Upper Myanmar. It is a grand festival during which local people honour spirits and pray for family health, a stable income and a good harvest (see Fig. 6.3).

During the festival, the shrine of the Shwe Kyun Pin Sister and Brother is surrounded by stalls selling local products such as bamboo crafts and unglazed water bottles, as well as local culinary specialities specific to the festival. Women from Ye Le Kyun and neighbouring villages sell seeds of various crops (see Fig. 6.4).

Farmers in flood-prone areas, where it is difficult to harvest or store seeds, purchase seeds even at relatively high prices at the festival. Each of the seed dealers who gather for the festivities sells on average seeds of 21 crop species, including *Hibiscus sabdariffa* (roselle) and *H. cannabinus* (kenaf), *Lablab purpureus* (hyacinth bean), *Vigna unguiculata* (cowpea, also known as black-eye pea), *Benincasa hispida* (wax gourd), *Ipomoea aquatica* (water spinach), *Luffa cylindrica* (sponge gourd), *Cucurbita maxima* (pumpkin) and *Coriandrum sativum* (coriander). Of these, *H. sabdariffa* and *H. cannabinus* are sold in particularly large quantities. According to tradition, seeds purchased at the festival are believed to promise good luck, and to yield a good harvest.

At the festival, some of the seeds produced in very small quantities by small-scale farmers are collected and mixed, and sold not only to locals, but also to farmers visiting from beyond the region. These seeds may include morphologically different phenotypes that we have seen in the village fields surveyed. In this way, diverse genetic resources are widely spread and shared throughout the festival. At the same time, farmers from Ye Le Kyun, whose conditions and location constrain seed production, ensure the dissemination of diverse strains of chinbaung via festival traditions.

Table 6.2 Relationship between seed production and distribution in three villages

Village name	Village location (and altitude)	Period of the field's availability	Crops cultivated	Number of seed producing farmers/Total number of farmers	Seed source	Relationship with Poe Yon	Shwe Kyun Pin Nat Festival
Ye Le Kyun	Riverbed and towhead of the Ayeyarwady river (65 m)	November–June (the fields are submerged during the rainy season from late June to October)	Vegetables Legumes	30–40/290	Seed saving by self Purchase at the Shwe Kyun Pin Nat Festival		Sell various seeds including chinbaung in small quantities for high prices
Yin Taw	River banks and **hinterland** of the Ayeyarwady River (74 m)	November–June (the fields are submerged during the rainy season from late June to October)	Vegetables Legumes	0/240	Purchase from seed farmers in Ywa Thit Purchase at the Shwe Kyun Pin Nat Festival (for growing in domestic gardens)	Sell harvested beans	Purchase a small amount of seeds for growing in domestic gardens
Ywa Thit	Plateau, distant from the Ayeyarwardy River (136 m)	Year-round	Vegetables Legumes Rice (+ Livestock)	100/330	Seed saving by self Seed shops Purchase at the Shwe Kyun Pin Nat Festival	Sell harvested beans and surplus vegetable and chinbaung seeds	Purchase a small amount of seeds for growing in domestic gardens

Fig. 6.3   A monk buying seeds at the Shwe Kyun Pin Nat Festival

Fig. 6.4   A seed seller at the Shwe Kyun Pin Nat Festival

6.6   THE ROLE OF POE YON,
A DISTINCTIVE STAKEHOLDER

Poe Yon is a kind of guild for the private agricultural brokerage firms seen in large and small cities in Myanmar. Each Poe Yon firm deals with crops, such as oil crops, rice, fresh vegetables, fruits or flowers. All of the firms are privately owned and operated, mainly by family members, and are connected with other Poe Yon firms in the same city and adjacent areas, forming a cooperative association for the exchange of information and regulation of commodity prices. They are also connected with their counterparts in other cities through broad information networks and distribution channels, and frequently use them to exchange goods directly or through bigger merchants, mainly from China.

In Ye Le Kyun, there are also farmers who sell chinbaung seeds alongside harvested legumes to Poe Yon firms run by relatives in Sagaing City. One Poe Yon owner from Ye Le Kyun reported that he sells most of the collected chinbaung seeds to dealers in Sagaing and Mandalay, and the rest to oil factories. We found that chinbaung was, unlike other kinds of vegetables, always much in demand by farmers and seed dealers both as seeds for vegetable production and as oilseed. This indicates the distinctive position occupied by chinbaung in Myanmar. Seeds from large-scale and small-scale farmers are also likely to be distributed via the Poe Yon network to various parts of Myanmar, thereby contributing to the maintenance of chinbaung genetic diversity as a whole.

The Shwe Kyun Pin Nat Festival and the unique relationship between Poe Yon firms and farmers play a critical role in the distribution of chinbaung seeds, and are very different from the distribution channels where major seed companies mediate with stakeholders, as is the case with improved varieties of vegetables (Fig. 6.5).

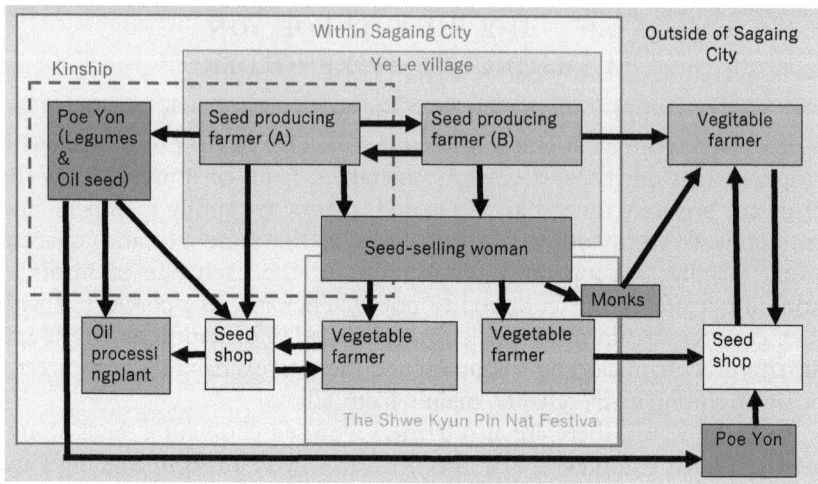

Fig. 6.5 Seed distribution channels in Sagaing City and outside of Sagaing City

6.7   Conclusion

Farmers in Myanmar use chinbaung in diverse ways. Many produce chinbaung seeds on-farm for the next cropping season. The main aim, aside from reducing costs, is to grow varieties with superior traits selected by the farmers themselves. However, large-scale rice farmers in the Sagaing and Mandalay Regions purchase seeds for more effective use of their plots via crop rotation, alternating chinbaung cultivation for vegetables with paddy rice production.

Our field survey revealed that small-scale farmers producing vegetables continue to grow diverse species and varieties of chinbaung that are developed and maintained not only according to constraints of the local environment, but also to the demands of livelihoods, through factors such as usefulness for subsistence, taste and versatility. Many traditional varieties are stored and managed in farmers' homes, so farmers who need seeds because of loss from flooding or drought, for instance, can obtain seeds similar to those of their neighbours and home villages. At the same time, the introduction of new varieties or genetic characteristics may lead to the diversification of local varieties.

From interviews with local farmers and members of Poe Yon, we noted that farmers' networks are intimate and cooperative, based on familial or territorial ties. In one Poe Yon firm we studied, the owner's wife was from Ye Le Kyun, thus her seed transactions in the village mainly took place with farmers she was related to. Each Poe Yon firm deals with large quantities of a limited number of items—a response to demand from neighbours and distant consumers, and based on a solid network with other Poe Yon firms across different cities. In the Mandalay Region, the Poe Yon firms have formed an association that sets crop prices. A similar association emerging in Yangon and other large cities in Myanmar coordinates its own seed-trading activities and even those with neighbouring China.

This is a good example of how a broad and continuous supply of diverse traditional seeds can be realised through farmers' own decisions, enacted in their fields and through their seed networks. Although farmers have apparently developed their seed acquisition and provision practices spontaneously, and although these practices appeared resilient during our survey, these practices cannot be seen as secure, given the advance of industrialised, commercialised farming in Myanmar. Yet seed diversity is an important component of biocultural heritage as well as sustainable agriculture. We recommend that farmers' practices be integrated into agricultural policy. Such a move contributes to the conservation of crop diversity in relatively small areas, and the sustainability of the human-made distribution system that facilitates crop diversity in larger ones.

Acknowledgements The research has been funded partially by JSPS KAKENHI 17H01682:17H04627.

References

Domon, E., Thein, M. S., Takei, E., Osada, T., & Kawase, M. (2015). A field study collecting cultivated crops and useful plants in Sagaing region of Myanmar in 2014. *Annual Report on Exploration and Introduction of Plant Genetic Resources, 31*, 343–365.

Ministry of Agriculture and Irrigation. (2015). *Myanmar rice sector development strategy: Building the foundation of a modern, industrialized nation through inclusive agricultural and rural development.* Retrieved July 21, 2021, from http://books.irri.org/MRSDS_content.pdf

Mohamed, B. B., Muhammad, B. S., Hassan, S., Rashid, B., Beenish, A., & Husnain, T. (2015). Tolerance of Roselle (*Hibiscus sabdariffa* L.): Genotypes to drought stress at vegetative stage. *Advancements in Life Sciences,* 2(2), 74–82.

Nagashima, M., Irie, K., Yoshida, S., Kikuno, H., Mar Saw, O., Soe, T. T., & Watanabe, K. (2019). Evaluation of diversity of plant genetic resources grown in Myanmar home garden: Distribution and utilization of *Hibiscus* genus plant "chinbao." *Journal of the International Society for Southeast Asian Agricultural Sciences,* 25(1), 104–111.

Nagashima, M., Yoshida, S., Kikuno, H., Wakui, K., Nishikawa, Y., Mar Saw, O., Moe, S., & Irie, K. (2019). Field survey and collection of "chinbao", *Hibiscus* spp. in Chin State of Myanmar (20th of December 2017–1st of January 2018). *Annual Report on Exploration and Introduction of Plant Genetic Resources,* 34, 137–146. https://doi.org/10.24514/00001138

Sharma, H. K., Sarkar, M., Choudhary, S. B., Kumar, A. A., Maruthi, R. T., Mitra, J., & Karmakar, P. G. (2016). Diversity analysis based on agro-morphological traits and microsatellite based markers in global germplasm collections of roselle (*Hibiscus sabdariffa* L.). *Industrial Crops and Products,* 89, 303–315. https://doi.org/10.1016/j.indcrop.2016.05.027

Thein, M. S., Kawase, M., Domon, E., & Watanabe, K. (2017). A field study to explore plant genetic resources in the Sagaing Region of Myanmar in 2015. *Annual Report on Exploration and Introduction of Plant Genetic Resources,* 33, 239–263. https://doi.org/10.24514/00001103

Tun, T. & Than, M. (1996). *Myanmar: Country report to the FAO international technical conference on plant genetic resources.* FAO. Retrieved July 21, 2021, from http://www.fao.org/fileadmin/templates/agphome/docume nts/PGR/SoW1/asia/MYANMAR.pdf

Organizations and Functions for Seed Management in East Asia: Korea, Japan and Taiwan

Mitsuyuki Tomiyoshi

Abstract How important are informal seed-saving systems in conserving agrobiodiversity? Mitsuyuki Tomiyoshi probes that question in the East Asian context in this survey and analysis examining the prevalence of community seed banks and other non-profits in Japan, South Korea and Taiwan. In Japan, traditional varieties are generally cultivated on the basis of heritage and culture, and an array of non-profits are involved in seed provision, domestic production, collection and networking. In South Korea, where interest in heirloom seeds has been growing since the early 2000s, comparable organizations include a research firm, cooperative and civil society network. Relevant non-profits in Taiwan, meanwhile, are at the formative stage. Informal systems are key to maintaining agrobiodiversity, Tomiyoshi concludes, but to operate sustainably they must better integrate their functions and set strategies for collaboration with public institutions.

M. Tomiyoshi (✉)
Kurume University, Kurume, Japan
e-mail: tomiyoshi_mitsuyuki@kurume-u.ac.jp

© The Author(s) 2022
Y. Nishikawa and M. Pimbert (eds.), *Seeds for Diversity and Inclusion*,
https://doi.org/10.1007/978-3-030-89405-4_7

Keywords Community seed bank (CSB) · Cultural heritage · Farmers' rights · Non-profit organization (NPOs) · Social movement

7.1   INTRODUCTION

In the field of seed management research, systems for saving seeds are classified as either formal (established by governments or seed companies) or informal (established by farmers producing or sharing seeds themselves). Various studies have concluded that the latter systems—networks of farmers sharing seeds—play a passive role, merely supplementing the functions of formal systems. However, this is a mistaken view.

Research has revealed the importance of, and diverse roles played by, farmers' seed networks (Coomes et al., 2015). One example of such an heirloom seed management system is the community seed bank (CSB). These organizations (also known as seed libraries or seed savers' networks) are generally expected to function as hubs in the region, promoting an effective means of conserving agricultural biodiversity.

Surveys of the activities of local CSBs are being conducted around the world. The banks have three notable functions: preserving crop genetic resources, promoting access to and use of a region's diverse crops and protecting rights to seeds and food (Vernooy et al., 2014). In Britain and the United States, many surveys of CSBs that function at the domestic level have been published (Curry, 2019; Helicke, 2015). Numerous studies have also examined their role in protecting farmers' rights (Vernooy et al., 2020).

In Japan, it is primarily small and family farmers who grow heirloom varieties. There is limited data on the conservation of heirloom cultivars by such farmers. The Japan Organic Agriculture Association (JOAA), a non-profit organization or NPO, conducted a study of on-farm seed saving by organic farmers in Japan, and found that 58.7% of such farmers are involved in the practice. Among Japanese farmers as a whole, however, organic farmers represent a very small proportion.

In addition, Mitsuyuki Tomiyoshi et al. (2021) investigated the number of farmers in the Noto Peninsula—a disadvantaged rural area of Japan with many small-scale farmers—to gauge the prevalence of seed collection. Among all locals growing some kind of crop, including full-time, part-time and subsistence farmers, some 21% were found to engage in the practice.

In Korea and Taiwan, numbers of smallholders that play a central role in the conservation and use of heirloom cultivars is declining because of demographic ageing. This is problematic, given the need to forge a relevant management system that involves a number of organizations. However, only a limited number of studies about seed conservation activities have been carried out in East Asia.

In this chapter, we will analyse the results of surveys about Japan, Korea and Taiwan to determine the roles CSBs and other NPOs play in East Asian seed conservation. Using this analysis, we will discuss the state of the practice in East Asia and compare functions of seed management by CSBs with those proposed by Ronnie Vernooy et al. (2014). We will also extend that framework to make it more universally applicable.

7.2   FUNCTION OF NON-PROFITS IN SEED CONSERVATION

NPOs are generally involved in the management of genetic resources through self seed saving, seed collection and storage, seed provision, food processing and sale and networking. As shown in Table 7.1, each of these activities has been carried out by a range of entities. NPOs, on the other hand, cover many of the activities simultaneously.

Table 7.1   The activities of entities involved in seed conservation and management

Function	Activities	Existing entities
Home seed production	Actively cultivate and produce seeds	Farmers
Seed collection and conservation	Collect and store varieties that are disappearing from the area	Gene banks/universities
Seed provision	Sell seed as a business	Seedling sellers
Food processing and sale	Process/sell harvested produce as local specialty products	Food businesses, etc.
Networking	Link various organizations	NPOs, etc.

7.3 SEED-CONSERVATION NON-PROFITS IN JAPAN

Fact-finding surveys were conducted focusing on four NPOs involved in the conservation and use of heirloom cultivars: the JOAA (registered under Japan's NPO Act; Kiyosumi no Mura (also NPO-registered); the Hiroshima Prefecture Agriculture and Forestry Promotion Centre Agriculture Seedbank, or Hiroshima Seedbank (a general incorporated foundation); and the Hyogo Heirloom Cultivar Conservation Association, or HCA (an informal civil organization not registered as an NPO).

Based on the survey of these bodies, and the results of the five categories in which they operate, a number of observations were made (see Table 7.2).

7.3.1 Home Seed Production

In each of the four organizations surveyed, association members (or staff) cultivated and produced the seeds of heirloom cultivars. In the case of Kiyosumi no Mura, approximately 150 cultivars are conserved on the farm. The JOAA and the HCA had implemented indirect frameworks for organizing home seed production. However, this was not done at an organizational level, but by members individually.

Table 7.2 A comparison of the conservation and use of heirloom cultivars by non-profits and other entities in Japan

Function	JOAA	Kiyosumi no Mura	Hiroshima Seedbank	HCA
Home seed production	Indirectly engaged[*]	Directly engaged	Directly engaged	Indirectly engaged[*]
Seed collection and conservation	Directly engaged	Directly engaged	Systematically engaged	Directly engaged
Seed provision	Indirectly engaged[**]	Indirectly engaged[**]	Seed loan programme[**]	N/A
Food processing and sale	N/A	Served in farmers' restaurant[*]	N/A	N/A
Networking	Nationwide	Local	N/A	Prefecture-wide

[*]Indicates the activities of closely related organizations/individuals
[**]Including distribution at no charge and seed exchange groups

7.3.2   Seed Collection and Conservation

Hiroshima Seedbank collects heirloom cultivars throughout the prefecture. Approximately 18,000 samples, including seeds and seedlings from outside the prefecture and from Japanese universities, are collected. (This seems a challenging task for a single prefecture, given that the NARO Genebank, a government agency, stores 248,000 samples.)

Rather than collecting heirloom cultivars as an organization, the JOAA mainly stores seeds and seedlings exchanged among its members. However, as in the case of similar organizations in other countries, it is possible that if seed swaps and other activities are held, farmers will come and offer seeds of crops they can no longer cultivate themselves. That would contribute significantly to cultivar conservation.

The HCA has embarked on a project to find heirloom cultivars grown on a small scale in various parts of Hyogo Prefecture. Rather than storing the seeds of the various varieties it finds, the association collects information about the farmers holding them and makes this information available to its members to encourage sustainable cultivation of the cultivars within the prefecture.

7.3.3   Seed Provision

The JOAA has plans to establish a seed and seedling sales division. But because of the restrictions in Japan's Plant Variety Protection and Seed Act (revised and approved in 2020) and the costs associated with selling such products, it would be difficult for an NPO to engage in it as a business.

7.3.4   Food Processing and Sale

None of the organizations surveyed was involved in food processing or selling (including the sale of agricultural produce), since such enterprises generally involve profit. However, Kiyosumi no Mura has established a system where all agricultural produce grown by members of a related village farm cooperative is bought by the farmers' restaurants managed by the leader of that NPO.

7.3.5   Networking

The JOAA has a nationwide network and promotes crop diversity by enabling members to share seeds and information. Kiyosumi no Mura mainly strengthens existing farmers' networks by cooperating with village authorities in Nara Prefecture. However, many artists, students and researchers from outside the prefecture participate in its activities, so it appears that the NPO also acts as a platform for connecting villages with stakeholders outside the locality. Painters, for example, have created calendars with pictures of traditional vegetables, and seed-themed events are sometimes held in the restaurants. HCA promotes the conservation and consumption of heirloom cultivars by connecting farmers, gardeners and consumers, mainly in Hyogo Prefecture.

7.4   Non-Profits and Similar Organizations in South Korea

The roles of South Korean organizations involved in the management of traditional cultivars, as indicated by a survey of three bodies operating in Korea, are presented in Table 7.3.

Each of the three organizations is engaged in home seed production aimed at cultivar renewal. Heuksalim, a joint stock company and related private research institute, conducts cultivation and seed production at its laboratory. Shinlim, an agricultural cooperative, preserves seed lines in its

Table 7.3 A comparison of private South Korean organizations and agricultural cooperatives involved in seed conservation

Function	Heuksalim	Shinlim agricultural cooperatives	Seedream
Home seed production	Directly engaged	Directly engaged	Directly engaged
Seed collection and conservation	Directly engaged	Directly engaged	Directly engaged
Seed provision	Indirectly engaged[*]	Indirectly engaged[*]	Indirectly engaged[*]
Food processing and sale	Directly engaged	Directly engaged	None
Networking	Directly engaged	Regional	National scale[**]

[*]Including distribution of seed gratis
[**]Network centred around Seedream founder and government research officer An Wan-Shok

fields. Seedream explores traditional varieties and also provides training for seed savers. Each of the organizations managed over 1000 strains.

The three bodies are on the whole larger than comparable NPOs and other organizations in Japan, a difference that may be partly down to their food processing and sales enterprises (described in Sect. 7.6). Because their main businesses are dealing primarily in traditional cultivars, it is possible for them, especially Heuksalim and Shinlim, to engage in the conservation of traditional cultivars—an expensive operation for a non-profit. Those two organizations launched in response to two big shocks that hit the South Korean economy within the past several decades: an intervention by the International Monetary Fund in 1997, and a Free Trade Agreement with the United States in 2008. Both of these initiatives had negative effects on the country's food sovereignty.

All three of the surveyed organizations are also engaged in seed collection and storage. At Seedream, seeds are managed in small storage buildings beside fields—a method also used by Japanese NPOs. However, we think that securing funds to continuously manage such a system is a major challenge if the activity is to continue.

Heuksalim's research centre provides cereal seed at no cost to its contract growers. Shinlim does something similar. However, since Seedream does not independently sell seeds of traditional cultivars, its function as a seed provider is limited. Overall, seed provision in South Korea is very different from that in Japan, where there are many small-scale seed and seedling retailers (such as Noguchi Seeds) that sell traditional, open-pollinated cultivars.

Heuksalim and Shinlim are also both engaged in food processing and selling, including the sale of agricultural produce. Both have commercialized operations by branding traditional varieties, which are not suitable for distribution as F1 hybrids are; and both provide vegetable delivery services to members. Japanese NPOs' efforts, by contrast, tend to be limited to conservation and rarely expand into sales and distribution.

Through their efforts to manage traditional cultivars with contract growers and other parties, Heuksalim and Shinlim have created networks. To date, these remain mostly localized. Seedream, however, has members in many regions and acts as a platform linking a wide variety of people. In fact, Seedream has more than 6000 members, making it the biggest network and technical association of the three organizations studied (Kim, 2013).

Seedream has also established relationships with both governmental and non-governmental organizations through its extensive networks. Seedream's founder, An Wan-Shok, himself has strong connections with many different organizations and groups, and is a key figure in the field of traditional cultivar management. He was the government research officer in charge of the Korea Genebank, in the National Agrobiodiversity Center (NAC) that promoted South Korean agriculture during the time when the government was applying the Green Revolution approach. The South Korean facility also had strong ties with Japan's national gene bank.

However, after meeting leaders of the Korean Peasant Women Association, An Wan-Shok was introduced to the importance of local seed conservation through Indigenous knowledge, especially that held by female elders. He did not explicitly support a food sovereignty movement, but offers his knowledge and network to support the preservation of native seeds by civil society, while keeping strong ties with governmental institutions. We can see this as a case of an integrated endogenous development approach within the sovereignty frame, combined with a rather neutral stance on science and technology in the public sector.

7.5   Non-Profits in Taiwan

The survey of NPOs in Taiwan was conducted in tandem with the agronomist Warren Kuo, the foremost authority on the conservation of heirloom cultivars in Taiwan. According to Kuo, only a few farmers in Taiwan produce their own seeds. Although a relatively large number still cultivate local varieties of beans and cucumbers, most farmers buy seeds and seedlings. However, the custom of home seed production is still strong in some Indigenous villages in eastern Taiwan.

7.5.1   Warren Kuo and His Networks

In 2011, Warren Kuo held a conference with six Taiwanese NPOs. Subsequently, the NPOs launched workshops and began to teach farmers home seed production methods. This approach echoes participatory plant breeding, but has not been adopted by the Taiwanese government, and remains within the domain of research and farming.

7.5.2   *Hope Market*

Hope Market is a private organization in Taichung City comprised of more than 30 organic farmers who hold a farmer's market twice a month. In 2011, the organization launched an heirloom seed conservation project.

Hope Market focuses on Formosan and heirloom rice of the indica variety. Although its farmers acquired cultivars from Thailand in the market's first year of its operation, from the second year onwards they purchased local cultivars and selected plants from those. As part of this project, they have formulated a model in which farmers can save their own seeds at home. They also plan to run cultivar exchange workshops.

7.6   Comparison of the Three Countries and Regions

We compared the nature of NPOs' activities in three East Asian countries (see Table 7.4), including government institutions and seed companies for reference.

Our findings showed that NPOs in Taiwan are still in the early stages of development, and also that there was no evidence among them of any involvement in collecting and conserving heirloom varieties. Thus, compared with Japanese NPOs and heirloom cultivar conservation societies, Taiwan's are at the formative stage.

However, additional surveys carried out by Warren Kuo revealed that in 2013, NPOs in Taiwan developed a network known as Farmers' Conservation of Seeds. This network spread throughout the country, and is expected to remain active. By contrast, the government has made little effort towards participatory plant breeding, leaving it to researchers and farmers to spearhead such efforts. In Taiwan, university researchers are playing a leading role in such activities.

7.7   Conclusion

In Japan, South Korea and Taiwan, just a handful of organizations are involved in the social movement against global agribusiness. However, a few of these NPOs are also collaborating with global networks such as Vía Campesina, and their work demonstrates an awareness of farmers' rights.

Table 7.4 A comparison of genetic resource/heirloom seed conservation efforts in Taiwan, Japan and Korea

Type of organization	Taiwan	Japan	South Korea
Government institution	Plant genetic resources in gene banks: approximately 80,000 items • Government has a passive attitude regarding participatory plant breeding	Plant genetic resources in gene banks: approximately 220,000 items • Government/local government support the conservation of traditional vegetables	Plant genetic resources in gene banks: approximately 200,000 items
Seed/seedling company	Links with Japanese seed companies due to the territory's suitability as a place for growing seeds Small- and medium-sized seed companies aiming to popularize open-pollinated varieties and home seed production	A diminishing number of small- to medium-sized seed companies Seed companies specialising in heirloom seeds exist	Many large-scale seed companies bought out by multinational corporations State of small- to medium-sized seed companies is unclear
NPOs	Conservation activities commenced on the basis of advice from university researchers Collection and conservation programmes for local heirloom cultivars introduced Nationwide network launched	National, prefectural and local organizations active throughout the country Many organizations of different sizes	Surveys and collaboration on a national level by farmers' organizations Conservation and dissemination by public interest incorporated foundations in collaboration with companies Small number of large-scale organizations

Source Author's on-site surveys

Cases in Japan explicitly show that those involved in seed-saving activities, whether farmers or citizens, continue to cultivate traditional varieties on the basis of heritage and culture, guided by family traditions and culinary rationales. Thus, the motivation there often seems to be a micro-awareness, allied to locality and kinship, rather than global movements or rights-based approaches. Thus, the greater context for seed cultivation and saving in Japan is a concern to pass on heirloom cultivars and farming culture.

In South Korea, interest in traditional varieties of crops and vegetables has been growing since the early 2000s, and the Korean Women's Peasant Association has been working to develop a movement for food sovereignty. In 2007, Seedream was established through the initiative of civil society activism and Heuksalim (see Sect. 7.4 and Table 7.3). Byeong-Seon Yoon et al. (2013) and Hyo Jeong Kim (2013) have mentioned that women play an important role in the conservation of Indigenous species in South Korea. There is an emphasis on seeds themselves rather than the concept of farmers' rights in collaborations between Seedream, Heuksalim and women farmers; but that does not discount the value of these collaborations in elevating women's position in farming villages and households.

Taiwanese CSBs established and run by NPOs and similar institutions are still in the early stages of development.

In both Japan and South Korea, there are NPOs, agricultural associations and public institutions at the prefectural and municipal levels that are fulfilling the role of CSBs. In many cases, such activities are included in the multifaceted activities of CSBs, described by Ronnie Vernooy et al. (2014) as preserving crop genetic resources and promoting access to and use of a region's diverse crops. However, it is rare for NPOs in these three countries to be attuned to social movements for food sovereignty.

For organizations working as community seed banks to function more reliably and sustainably both in local communities and on a global scale, it is essential to integrate their different roles and functions, and also to establish strategies for co-existence and collaboration among NPOs and public institutions.

Acknowledgements This research was partly funded by the Japan Society for the Promotion of Science (JSPS) KAKENHI (Grant Number 24658194; 25850160; 16K18767; 20K06289).

REFERENCES

Coomes, O., McGuire, S., Garine, E., Caillon, S., McKey, D., Demeulenaere, E., et al. (2015). Farmer seed networks make a limited contribution to agriculture? Four common misconceptions. *Food Policy, 56*, 41–50. https://doi.org/10.1016/j.foodpol.2015.07.008

Curry, H. A. (2019). Gene banks, seed libraries, and vegetable sanctuaries: The cultivation and conservation of heritage vegetables in Britain, 1970–1985. *Culture, Agriculture, Food and Environment, 41*(2), 87–96. https://doi.org/10.1111/cuag.12239

Helicke, N. A. (2015). Seed exchange networks and food system resilience in the United States. *Journal of Environmental Studies and Sciences, 5*, 636–649. https://doi.org/10.1007/s13412-015-0346-5

Kim, H. J. (2013, September). *Women's indigenous knowledge and food sovereignty: Experiences from KWPA's movement in South Korea.* Paper for discussion at the Food Sovereignty: A Critical Dialogue international conference, New Haven, CT. Retrieved July 21, 2021, from https://www.tni.org/files/download/71_hyojeong_2013_0.pdf

Tomiyoshi, M., Uchiyama, Y., & Kohsaka, R. (2021). Evaluating plant genetic diversity maintained by local farmers and residents: A comprehensive assessment of continuous vegetable cultivation and seed-saving activities on a regional scale in Japan. *The International Journal of Sociology of Agriculture and Food, 26*, 2. https://doi.org/10.48416/ijsaf.v26i2.433

Vernooy, R., Sthapit, B., Galluzzi, G., & Shrestha, P. (2014). The multiple functions and services of community seedbanks. *Resources, 3*, 636–656. https://doi.org/10.3390/resources3040636

Vernooy, R., Mulesa, T. H., Gupta, A., Jony, J. A., Koffi, K. E., Mbozi, H., & Wakkumbure, C. L. K. (2020). The role of community seed banks in achieving farmers' rights. *Development in Practice, 30*(5), 561–574. https://doi.org/10.1080/09614524.2020.1727415

Yoon, B.-S., Song, W. K., & Lee, H. J. (2013). The struggle for food sovereignty in South Korea. *Monthly Review, 65*(1), 56–62. https://doi.org/10.14452/MR-065-01-2013-05_5

CHAPTER 8

Evolutionary Populations for Sustainable Food Security and Food Sovereignty

Salvatore Ceccarelli, Stefania Grando, Maedeh Salimi, and Khadija Razavi

Abstract Two mechanisms in plant breeding are thought to diminish crop diversity: the displacement of landraces by "improved" varieties, and a bias towards varieties developed under a high-input management regime. This multinational study examines how genetic diversity can be restored through evolutionary plant breeding: enabling plants under cultivation to evolve via natural selection pressure and adapt to the environment. The authors first present findings from research in Iran. Here, in participation with institutions, farmers selected barley, rice and wheat varieties from evolutionary populations for cultivation and used them outright as "smart crops" with all-around benefits for the environment, human health and farming income. A similarly successful project in Italy led to six more in countries across Africa, Asia and the Near East. Ultimately, the authors conclude, such "evolutionary-participatory"

S. Ceccarelli (✉) · S. Grando
Ascoli Piceno, Italy

M. Salimi · K. Razavi
CENESTA, Tehran, Iran

© The Author(s) 2022
Y. Nishikawa and M. Pimbert (eds.), *Seeds for Diversity and Inclusion*,
https://doi.org/10.1007/978-3-030-89405-4_8

plant breeding enables farmers to manage genetic diversity autonomously. While the seeds produced have yet to meet the requirements of seed laws, new rules emerging in Europe could enable organic farmers to adopt the approach from 2022.

Keywords Crop diversity · Evolutionary plant breeding · Iran · Italy · Seed autonomy

8.1   INTRODUCTION

Plant breeding is thought to be a cause of the decline in agrobiodiversity through two interconnected mechanisms: the displacement of myriad landraces by a few improved varieties (Dwivedi et al., 2016; van der Wouw et al., 2010), and the centralized organization of most breeding programmes in which selection takes place within one or a few well-managed research stations. The crops in the research stations are well managed (fertilized, protected against weeds, diseases and insects, and irrigated when necessary) because under those conditions selection is believed to be more efficient (Baranski, 2015; Ceccarelli, 1989). This belief, however, is not necessarily true.

Inevitably, the varieties produced by these breeding programmes can perform well if managed as they were during the selection process. This has two consequences: first, the dissemination of varieties will be accompanied by an increase in the use of inputs such as synthetic fertilizers and other chemicals; and secondly, as the inputs effectively minimize the differences between locations, the same varieties will be able to perform well widely, causing an overall decline of crop diversity (Bonnin et al., 2014).

The loss of agrobiodiversity is associated with increased vulnerability both to climate change (Keneni et al., 2012), including climate extremes (Isbell et al., 2015), and to agricultural pests (Díaz et al., 2006; Fisher et al., 2018). This situation is growing even more alarming because of the complexity of climate change, which involves not only a change in temperature and rainfall but also in the type and spread of insects, diseases and weeds (Ceccarelli & Grando, 2020a).

8.2   Bringing Back Diversity in Farmers' Fields: Participatory Plant Breeding

Participatory research has been formally proposed in the early 1980s (Rhoades & Booth, 1982) by social scientists, based on the principle of involving users and clients in research and development (Ceccarelli & Grando, 2020b). When applied to plant breeding, it was construed as a more socially equitable model. Along with its social dimension, participatory plant breeding (PPB) is also more efficient: its impact is measured not only as genetic gains and number of varieties released, but also as customer acceptance of the final product, a rise in agrobiodiversity and a higher benefit/cost ratio (Ceccarelli, 2015).

An example of farmers' assessment of the benefits of PPB was the International Farmers' Conference organized by the International Center for Agricultural Research in the Dry Areas (ICARDA) in May 2008, 13 years after PPB had started in Syria by the barley improvement program of the centre with the financial assistance of GTZ (German Corporation for Technical Cooperation) now GIZ. The conference involved more than 50 farmers from Algeria, Canada, Egypt, Eritrea, France, Iran, Jordan and Syria, and used different methodologies, including story-telling, to facilitate sharing their agricultural knowledge and thus capture their thoughts about PPB. The main findings (Galiè et al., 2009) suggest that farmers perceived that their participation in key decisional stages of plant breeding enhanced their self-esteem, increased their knowledge and communication skills, and changed their perception of gender roles.

PPB has been very successful among farmers' communities in Syria, Jordan, Yemen, Egypt, Tunisia, Morocco, Eritrea, Ethiopia and Iran (Ceccarelli et al., 2013) but not well accepted by several public research institutions, with a few exceptions. The reasons for this generalized institutional reluctance to adopt PPB have been discussed recently by Salvatore Ceccarelli and Stefania Grando (2020b). They range from conventional and biotechnological methods dominating university curricula on plant breeding to the reward system in public institutions, which is still largely based on the number of varieties released. A more fundamental reason is the reluctance to accept the paradigm shift that PPB inevitably implies in "seed sovereignty" and, consequently, "food sovereignty" (Ceccarelli & Grando, 2020b). In several countries, it has been reported that any institutional support was mostly of a personal nature and ended when the person left.

One notable aspect of the PPB programmes was the farmers' interest in experimenting by mixing different varieties of the same crop starting in 2007, particularly in Syria and Iran. Before the project, as with most farmers in the Near East, they were accustomed to receiving a few uniform varieties from the national research systems. In Syria, farmers' initial exposure to mixtures of different varieties was partly associated with the system of compensation for land use we developed in consultation with farmers. In fact, farmers agreed on a given amount of grain as fair compensation for the land and the time they dedicated to the experiments. Sometimes the grain was a mixture that, although intended as animal feed, was used at least in part as seed. It was in this way that farmers started to become aware of the advantages in growing mixtures such as better and more stable grain yield.

Syrian farmers also occasionally attended events at ICARDA's headquarters, where they had the opportunity to interact with the centre's top management. Yet the farmers' positive feedback failed to convince other scientists to follow the path of the centre's barley improvement programme. PPB never became ICARDA's main method for plant breeding.

8.3   From Participatory
to Evolutionary Plant Breeding

Although the term evolutionary plant breeding (EPB) was first used in 1956 (Suneson, 1956), the idea of exploiting the advantages of genetic heterogeneity, in the form of either mixtures, obtained by mixing the seed of different varieties of the same crop, or populations, obtained by mixing the seed derived from crossing different varieties of the same crop (Wolfe & Ceccarelli, 2019), was much older (Harlan & Martini, 1929, 1938).

After those early studies, a vast body of research has been published, demonstrating how evolutionary populations (EPs) and mixtures are able to evolve and adapt their phenology to the area of cultivation by becoming earlier maturing or later maturing depending on the adaptative advantages of these traits (Allard & Hansche, 1964), to increase their yield (Patel et al., 1987; Rasmusson et al., 1967; Soliman & Allard, 1991; Suneson, 1956), their yield stability (Allard, 1961) and their height (Suneson & Wiebe, 1942).

A major benefit of EPs and mixtures is their ability to control the spread of diseases because the presence of plants with different levels of resistance and/or susceptibility makes the diffusion of the disease much slower than in a genetically uniform crop (Finckh & Wolfe, 2006; Finckh et al., 2000; Ibrahim & Barret, 1991; McDonald et al., 1988; Mulumba et al., 2012; Mundt, 2002; Simmonds, 1962; Smithson & Lenné, 1996).

More recently, several papers confirmed that EPs do adapt to different geographical areas via phenological adaptation (Goldringer et al., 2006), that they tend to perform better than uniform varieties in years affected by drought (Danquah & Barrett, 2002) and that they can combine higher yield and higher yield stability (Raggi et al., 2017). Eventually, the evolutionary potential of EPs and mixtures, if widely adopted, can represent a fast and economic solution to the complexity of climate change (Ceccarelli & Grando, 2020a). It is therefore surprising that, despite all the scientific evidence, there has been very little practical agricultural use of EPs beyond their discussion in scientific papers.

Because of the difficulties in institutionalizing PPB, in 2008, while working at ICARDA, two of us (S.C. and S.G.) decided to constitute an EP of barley (*Hordeum vulgare* L.) by mixing the seed of the progenies of 1600 crosses. This EP was then planted in Syria, Eritrea, Jordan, Algeria and Iran—some of the countries where a network of farmers had already enhanced their diversity management skills through participation in PPB programmes. While two additional EPs, one with bread wheat (*Triticum aestivum* L.) and one with durum wheat (*Triticum turgidum* L. subsp. *durum* [Desf.] Husn.) were developed at ICARDA, the barley EP planted in Iran by two farmers in Kermanshah and Semnan provinces caught the attention of an Iranian breeder at the Dryland Agricultural Research Institute (DARSI) in Kermanshah, who decided to make a local bread wheat population.

8.4 The Evolutionary Populations in Iran

Iranian farmers who planted the barley EP were so satisfied with the population's performance that they shared the seed with farmers in other provinces, both through the PPB programme of Iran's Centre for Sustainable Development and Environment (CENESTA), and informally with neighbours, friends and relatives. This initial work with EPs in Iran by CENESTA, was supported by a small grant from the International Fund for Agricultural Development (IFAD) and within a few years, from 2010

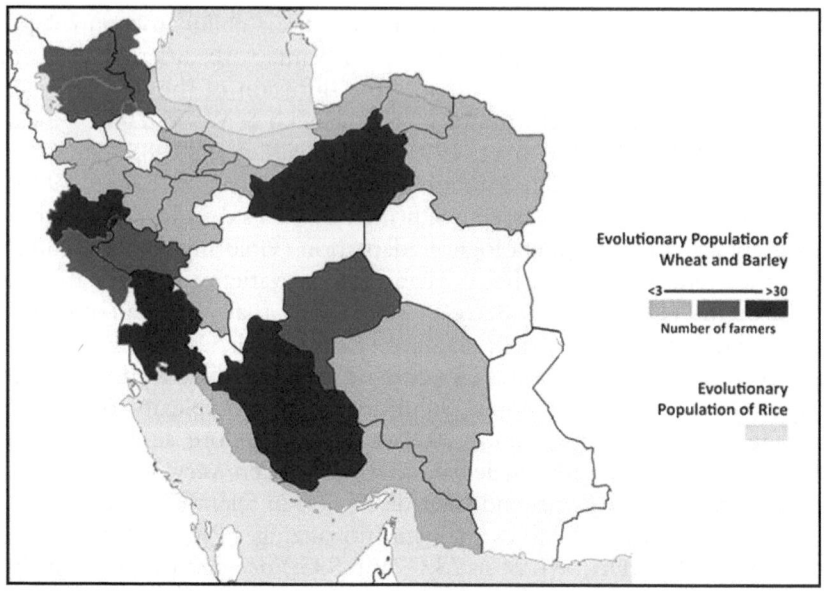

Fig. 8.1 Distribution of wheat, barley and rice evolutionary populations in Iran

to 2014, the populations covered several hundred hectares in 17 provinces and involved some 150 farmers (see Fig. 8.1).

EPs in Iran are mostly cultivated under rainfed conditions. But they are grown, too, by farmers in irrigated areas who are facing water shortages, because the populations are recognized as more resistant to water scarcity and drought than modern varieties.

Indeed, most of the Iranian farmers in marginal[1] areas and under low-input and rainfed conditions reported that EPs had higher yields than modern varieties and showed good yield stability. They are more resistant to biotic and abiotic stresses such as pests and diseases, weeds, drought, water shortage cold, heat, strong winds and hail, and can grow better than uniform varieties in low-fertility soils. In addition, many farmers reported the EP seeds as very beautiful and, in terms of quality, health and thousand-kernel weight, superior to local, improved or modern varieties.

[1] Marginal here is used to mean poor soil, low rainfall with low expected agricultural productivity, but also socially marginal with limited opportunity for other jobs.

There are also reports on food and feed quality in these populations. For example, baking bread with flour from EPs has been found to improve its smell and taste, as well as nutritional qualities. In addition, nomads or herders have noted that as livestock feed, the grains or straw of the ICARDA barley EP, compared with conventional feed, accelerates growth in lambs and improves the quality of milk. All of these characteristics encourage farmers to use and multiply these populations and cultivate them regularly as a main crop.

Our original aim regarding EPs was to endow farmers with a wealth of genetic diversity from which they could select, independently in each location, varieties adapted to their physical, social and market conditions. One unexpected outcome of the IFAD project, however, was that the bread obtained from the bread wheat EP—cultivated without any further selection—soon became the basis of a profitable business. That opened up the possibility of using the populations as crops. There were a number of advantages to this approach: it exploited the evolutionary ability of the population to slowly adapt to both short- and long-term climate change with the complexity described earlier; it enabled the development of an independent seed supply, because there is no better seed than the one that continuously adapts to farmer's conditions; and it generated income.

In the EPB programme in Iran, the EPs reach farmers through various stakeholders, but in most cases, the seeds are distributed through a farmer-saved seed system. CENESTA had and continues to have a facilitating role in ensuring that farmers, breeders and researchers can connect, and also ensures that EPs are steadily disseminated among groups such as farmers, research stations and extension centres.

In the first year under the programme, each farmer receives 4–10 kilograms of one of the EPs, depending on the amount of seed available and the severity of environmental and climatic conditions in different regions. Farmers in more environmentally stressed and less productive areas are given more seeds; thus, if they lose most of the genotypes within the EP, they have a higher probability that a sufficient number survive from which to collect spikes for next year's cultivation. In the following year, and if farmers are satisfied with the EP's results, they will start seed multiplication to produce enough seeds and to expand the area under cultivation. On average, after four years, the farmers who had the most success with the crop and consequently with more seed availability will start exchanging the seed of the EPs, even up to several tons, with other farmers in their region or those with similar environmental and climatic conditions (see Fig. 8.2).

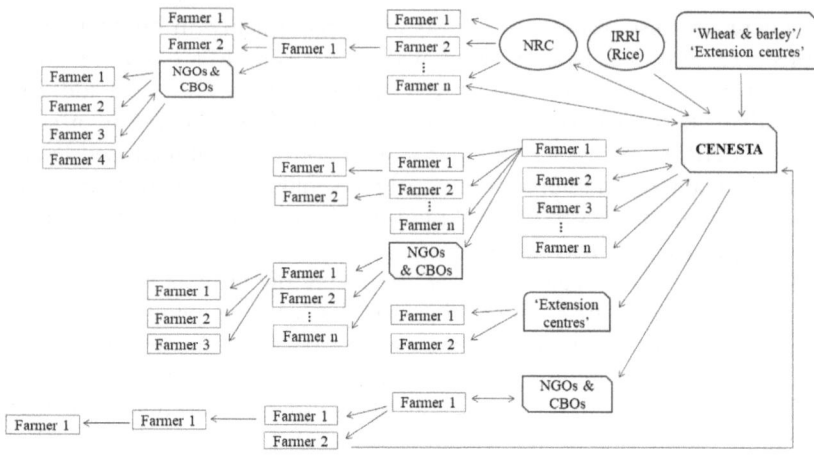

Fig. 8.2 Farmer to farmer diffusion of evolutionary populations (EPs) in Iran: the barley EP originated from ICARDA, the bread wheat EP originated from DARSI and the EP of rice originated from Iranian landraces received from IRRI. The EPs were initially distributed by CENESTA either directly to farmers or through other organizations (NRC = National Research Centre, NGOs = Non-Governmental Organizations, CSOs = Civil Society Organizations)

In recent years, both the number of farmers and the area under cultivation with EPs have increased, although monitoring and estimating that increase is not easy, given the complexity and breadth of farm-saved seed systems. It is known, however, that hundreds of farmers in different parts of Iran do cultivate EPs and extend the area under cultivation every year, and also exchange their seed with other farmers in their region.

The success of EPB in Iran with wheat and barley suggested to CENESTA researchers that the same approach might be used with rice, one of the country's most important food crops. Beginning in 2013, CENESTA imported 210 Iranian rice landraces from the gene bank of the International Rice Research Institute (IRRI) and multiplied them at the Research Station of the Rice Research Iranian Institute (RRII). Three evolutionary mixtures were then developed, by mixing the seed of all the landraces (200 genotypes), only the early maturing varieties (115 genotypes) and the late-maturing varieties (85 genotypes). These were distributed to some farmers and were also used in a project between RRII and CENESTA and, in 2019, in a new IFAD-funded project (see Sect. 8.6).

8.5   Smart Food from Iran's Evolutionary Populations

EPs may be considered to be smart crops—healthy for the consumers, as well as produced and distributed in an environmentally sustainable way and profitable for the farmers.

The benefits of EPs for consumers are manifold. The bread made using the first EP of bread wheat grown in Iran was highly appreciated by consumers for its digestibility, lengthy shelf life, flavour and aroma. Similar developments took place in Italy beginning in 2010 (see Sect. 8.6).

Regarding EPs' benefits for the environment, their diversity renders them resistant to diseases, insects and weeds, making the use of pesticides unnecessary; that in turn reduces emissions and helps to mitigate the effects of climate change. As the EPs evolve, they also adapt to the unpredictable and location-specific complexities of climate change.

EPs are good, too, for the farmers who grow them. Their robustness and resilience reduce production costs, while the popularity of the products derived from the different wheat EPs brings in income.

Even as the cultivation of EPs increases, and the products made with them reach more organic shops, researchers are still drawn to the science of populations (Ceccarelli & Grando, 2020a; Raggi et al., 2017). The scientific literature shows how populations evolve by becoming more productive and disease-resistant, how they ripen in harmony with the environment in which they evolve and how their yields become more stable year on year.

8.6   Evolutionary Populations in Other Countries

The success of EPs in Iran had two major international consequences. The first was in Italy, where the three ICARDA EPs arrived in 2010 via the Italian Association of Organic Agriculture, and quickly spread to farms throughout the country. The Italian experience of the bread wheat EP mirrored Iran's, and in a few years farmers in most regions of Italy (Fig. 8.3) were growing the EP, millers were selling the flour, bakeries were producing much-appreciated bread from it and organic shops and restaurants were selling both bread and flour. The same happened with a durum wheat EP, grown for use in pasta, although not yet as widely as the bread wheat EP.

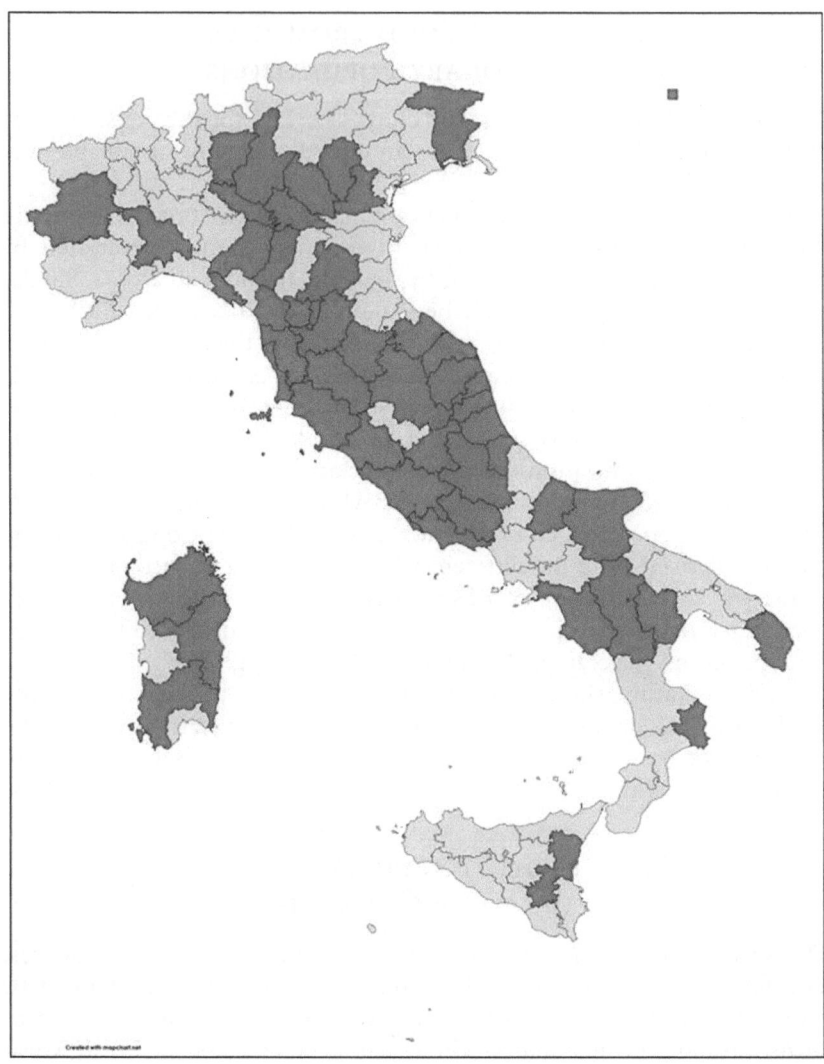

Fig. 8.3 The spread of evolutionary populations of bread wheat, durum wheat and barley in Italy from 2010 to present

The second consequence was that IFAD, following the success of its small grant project (see Sect. 8.4), decided to invest in evolutionary plant breeding and financed a four-year (2018–2022) project in Africa (Uganda and Ethiopia), the Near East (Jordan and Iran) and Asia (Nepal and Bhutan) implemented by Bioversity International. The project covers important staple crops such as wheat (Ethiopia, Jordan and Iran), rice (Iran, Nepal and Bhutan), bean (Uganda, Nepal and Bhutan) and barley (Ethiopia, Jordan and Iran). At the time of writing, the project has ended its third year's trials planted in 46 locations across the 6 countries with over 450 EPs, using local and improved varieties as controls.

8.7   Conclusions

Evolutionary-participatory plant breeding can be considered as a further step, compared with PPB, towards what Jack Kloppenburg (2010) has defined as "repossession". In fact, evolutionary-participatory plant breeding allows farmers to autonomously manage genetic diversity without institutional support. This does not mean that Research institutions are excluded from the process; only that their participation is not indispensable. Institutions may play a role in developing populations by crossing a given number of varieties or simply making available to farmers the remnant seed of early segregating populations. Beyond providing farmers with the diversity offered by EPs, institutions may develop a decentralized-participatory breeding programme distributing EPs assembled by them to farmers representing the target population of environments the programme aims to serve as, for example, farmers practising organic agriculture in geographical areas characterized by different rainfall, elevation, disease pressure and weed but also different market opportunities (Ceccarelli & Grando, 2020c). Under such a programme, farmers conduct the selection process in their own fields, and institutions multiply the selected material and organize field trials.

The examples of Iran, Italy and other countries indicate that EPs can be defined as "smart crops" because they represent a triple "win". They are good for the planet, as they reduce the use of chemical inputs and allow adaptation to the complexity of climate change; they are good for the consumer as they produce healthy food and they are good for farmers as they generate income.

The main hurdle in the diffusion of EPs is seed laws. EPs do not meet the Distinctness, Uniformity and Stability (DUS) requirements for their official registration and marketing, established by the International Union for the Protection of New Varieties of Plants (UPOV). However, the European Union has recently taken two interesting steps.

First, a Commission Implementing Decision of 18 March 2014 pursuant to Council Directive 66/402/EEC in Europe made it possible to market experimentally heterogeneous materials, namely evolutionary populations and mixtures as defined earlier, of wheat, barley, oat and maize up to 28 February 2021. Secondly, a newly approved regulation of organic agriculture, to come into effect on 1 January 2022, will make it possible to use heterogeneous material in organic farming. That will not only enable organic farmers to use their own seed, but they will also be able to use EPs in decentralized-participatory breeding programmes to develop varieties and/or populations specifically adapted to organic agriculture, as described above.

Acknowledgements The authors thank IFAD for supporting the initial activities in Iran and for financing the project "Use of genetic diversity and Evolutionary Plant Breeding for enhanced farmer resilience to climate change, sustainable crop productivity, and nutrition under rainfed conditions in Uganda, Ethiopia, Jordan, Iran, Bhutan and Nepal". We also thank ICARDA for supporting the development of the initial EPs; IRRI for facilitating the introduction of Iranian rice landraces; Dr. Reza Hagparast for his support; Bioversity International for successfully implementing the IFAD project; the farmers in several provinces in Iran; and the CENESTA team (Mobina Nourmohamadian, Mehdi Esmaeli, Yashar Hassannejad, Hanie Moghani and Soheil Hosseinzadeh) for carrying out most of the fieldwork.

REFERENCES

Allard, R. W. (1961). Relationship between genetic diversity and consistency of performance in different environments. *Crop Science, 1*(2), 127–133. https://doi.org/10.2135/cropsci1961.0011183X000100020012x

Allard, R. W., & Hansche, P. E. (1964). Some parameters of population variability and their implications in plant breeding. *Advances in Agronomy, 16,* 281–325. https://doi.org/10.1016/S0065-2113(08)60027-9

Baranski, M. R. (2015). Wide adaptation of Green Revolution wheat: International roots and the Indian context of a new plant breeding ideal, 1960–1970. *Studies in History and Philosophy of Biological and Biomedical Sciences, 50,* 41–50. https://doi.org/10.1016/j.shpsc.2015.01.004

Bonnin, I., Bonneuil, C., Goffaux, R., Montalent, P., & Goldringer, I. (2014). Explaining the decrease in the genetic diversity of wheat in France over the 20th century. *Agriculture, Ecosystems & Environment, 195*, 183–192. https://doi.org/10.1016/j.agee.2014.06.003

Ceccarelli, S. (1989). Wide adaptation: How wide? *Euphytica, 40*, 197–205. https://doi.org/10.1007/BF00024512

Ceccarelli, S. (2015). Efficiency of plant breeding. *Crop Science, 55*, 87–97. https://doi.org/10.2135/cropsci2014.02.0158

Ceccarelli, S., & Grando, S. (2020a). Evolutionary plant breeding as a response to the complexity of climate change. *iScience, 23*(12), 101815. https://doi.org/10.1016/j.isci.2020.101815

Ceccarelli, S., & Grando, S. (2020b). Participatory plant breeding: Who did it, who does it and where? *Experimental Agriculture, 56*(1), 1–11. https://doi.org/10.1017/S0014479719000127

Ceccarelli, S., & Grando, S. (2020c). Organic agriculture and evolutionary populations to merge mitigation and adaptation strategies to fight climate change. *South Sustainability, 1*(2), e002. https://doi.org/10.21142/SS-0102-2020-013

Ceccarelli, S., Galiè, A., & Grando, S. (2013). Participatory breeding for climate change-related traits. In C. Kole (Ed.), *Genomics and breeding for climate-resilient crops* (Vol. 1, pp. 331–376). Springer-Verlag. https://doi.org/10.1007/978-3-642-37045-8_8

Danquah, E. Y., & Barrett, J. A. (2002). Grain yield in composite cross five of barley: Effects of natural selection. *Journal of Agricultural Science, 138*(2), 171–176. https://doi.org/10.1017/S0021859601001678

Díaz, S., Fargione, J., Chapin, F. S. III, & Tilman, D. (2006). Biodiversity loss threatens human well-being. *PLoS Biology, 4*(8), e277. https://doi.org/10.1371/journal.pbio.0040277

Dwivedi, S. L., Ceccarelli, S., Blair, W. M., Upadhyaya, H. D., Are, A. K., & Ortiz, R. (2016). Landrace germplasm for improving yield and abiotic stress adaptation. *Trends in Plant Science, 21*(1), 31–42. https://doi.org/10.1016/j.tplants.2015.10.012

Finckh, M. R., Gacek, E. S., Goyeau, H., Lannou, C., Merz, U., Mundt, C. C., Munk, L., Nadziak, J., Newton, A., de Vallavieille-Pope, C., & Wolfe, M. S. (2000). Cereal variety and species mixtures in practice, with emphasis on disease resistance. *Agronomy for Sustainable Development, 20*, 813–837. https://doi.org/10.1051/agro:2000177

Finckh, M. R., & Wolfe, M. S. (2006). Diversification strategies. In B. Cooke, D. Jones, & B. Kaye (Eds.), *The epidemiology of plant diseases* (pp. 269–307). Springer Netherlands. https://doi.org/10.1007/1-4020-4581-6

Fisher, M. C., Hawkins, N. J., Sanglard, D. M., & Gurr, S. J. (2018). Worldwide emergence of resistance to antifungal drugs challenges human health and food security. *Science, 360*(6390), 739–742. https://doi.org/10.1126/science.aap 7999

Galiè, A., Hack, B., Manning-Thomas, N., Pape-Christiansen, A., Grando, S., & Ceccarelli, S. (2009). Evaluating knowledge sharing in research: The International Farmers' Conference organized at ICARDA. *Knowledge Management for Development Journal, 5*(2), 108–126. https://doi.org/10.1080/194741 90903387666

Goldringer, I., Prouin, C., Rousset, M., Galic, N., & Bonnin, I. (2006). Rapid differentiation of experimental populations of wheat for heading time in response to local climatic conditions. *Annals of Botany, 98*(4), 805–817. https://doi.org/10.1093/aob/mcl160

Harlan, H. V., & Martini, M. L. (1929). A composite hybrid mixture. *Agronomy Journal, 21*(4), 487–490. https://doi.org/10.2134/agronj1929.000219620 02100040014x

Harlan, H. V., & Martini, M. L. (1938). The effect of natural selection in a mixture of barley varieties. *Journal of Agricultural Research, 57*(3), 189–199.

Ibrahim, K. M., & Barret, J. A. (1991). Evolution of mildew resistance in a hybrid bulk population of barley. *Heredity, 67,* 247–256. https://doi.org/10.1038/hdy.1991.86

Isbell, F., Craven, D., Connolly, J., Loreau, M., Schmid, B., Beierkuhnlein, C., … & Eisenhauer, N. (2015). Biodiversity increases the resistance of ecosystem productivity to climate extremes. *Nature, 526*(7574), 574–577. https://doi.org/10.1038/nature15374

Keneni, G., Bekele, E., Imtiaz, M., & Dagne, K. (2012). Genetic vulnerability of modern crop cultivars: Causes, mechanism and remedies. *International Journal of Plant Research, 2*(3), 69–79. https://doi.org/10.5923/j.plant.201 20203.05

Kloppenburg, J. (2010). Impeding dispossession, enabling repossession: Biological open source and the recovery of seed sovereignty. *Journal of Agrarian Change, 10*(3), 367–388. https://doi.org/10.1111/j.1471-0366.2010.002 75.x

McDonald, B. A., Allard, R. W., & Webster, R. K. (1988). Responses of two, three, and four component barley mixtures to a variable pathogen population. *Crop Science, 28*(3), 447–452. https://doi.org/10.2135/cropsci1988.0011183X002800030003x

Mulumba, J. W., Nankya, R., Adokorach, J., Kiwuka, D., Fadda, C., De Santis, P., & Jarvis, D. I. (2012). A risk-minimizing argument for traditional crop varietal diversity use to reduce pest and disease damage in agricultural ecosystem of Uganda. *Agriculture, Ecosystem and the Environment, 157,* 70–86. https://doi.org/10.1016/j.agee.2012.02.012

Mundt, C. C. (2002). Use of multiline cultivars and cultivar mixtures for disease management. *Annual Review of Phytopathology, 40,* 381–410. https://doi. org/10.1146/annurev.phyto.40.011402.113723

Patel, J. D., Reinbergs, E., Mather, D. E., Choo, T. M., & Sterling, J. D. E. (1987). Natural selection in a doubled-haploid mixture and a composite cross of barley. *Crop Science, 27*(3), 474–479. https://doi.org/10.2135/cropsc i1987.0011183X002700030010x

Raggi, L., Ciancaleoni, S., Torricelli, R., Terzi, V., Ceccarelli, S., & Negri, V. (2017). Evolutionary breeding for sustainable agriculture: Selection and multi-environment evaluation of barley populations and lines. *Field Crops Research, 204,* 76–88. https://doi.org/10.1016/j.fcr.2017.01.011

Rasmusson, D. C., Beard, B. H., & Johnson, F. K. (1967). Effect of natural selection on performance of a barley population. *Crop Science, 7,* 543–543. https://doi.org/10.2135/cropsci1967.0011183X000700050042x

Rhoades, R. E., & Booth, R. H. (1982). Farmer-back-to-farmer: A model for generating acceptable agricultural technology. *Agricultural Administration, 11,* 127–137. https://doi.org/10.1016/0309-586X(82)90056-5

Simmonds, N. W. (1962). Variability in crop plants, its use and conservation. *Biological Reviews, 37*(3), 422–465. https://doi.org/10.1111/j.1469-185X. 1962.tb01620.x

Smithson, J. B., & Lenné, J. M. (1996). Varietal mixtures: A viable strategy for sustainable productivity in subsistence agriculture. *Annals of Applied Biology, 128*(1), 127–158. https://doi.org/10.1111/j.1744-7348.1996.tb07096.x

Soliman, K. M., & Allard, R. W. (1991). Grain yield of composite cross populations of barley: Effects of natural selection. *Crop Science, 31*(3), 705–708. https://doi.org/10.2135/cropsci1991.0011183X003100030032x

Suneson, C. A. (1956). An evolutionary plant breeding method. *Agronomy Journal, 48*(4), 188–191. https://doi.org/10.2134/agronj1956.000219620 04800040012x

Suneson, C. A., & Wiebe, G. A. (1942). Survival of barley and wheat varieties in mixtures. *Journal of the Agronomy Society of America, 34*(11), 1052–1056. https://doi.org/10.2134/agronj1942.00021962003400110010x

van der Wouw, M., Kik, C., van Hintum, T., van Treuren, R., & Visser, B. (2010). Genetic erosion in crops: Concept, research results and challenges. *Plant Genetic Resources, 8*(1), 1–15. https://doi.org/10.1017/S14792621 09990062

Wolfe, M. S., & Ceccarelli, S. (2019). The increased use of diversity in cereal cropping requires more descriptive precision. *Journal of the Science of Food and Agriculture, 100*(11), 4119–4123. https://doi.org/10.1002/jsfa.9906

Adding Value to a Scottish Rye Landrace: Collaborative Research into New Artisanal Products

Stan Blackley, David McVey, Maria Scholten, and Adam Veitch

Abstract Hebridean rye (*Secale cereale*), a high-yield landrace grown by crofters in Scotland's Highlands and Islands, has traditionally been used as

The project is an output of Crofters Diversity Pays! (CDP), a partnership involving the Scottish Crofting Federation, Queen Margaret University, Edinburgh and Science and Advice for Scottish Agriculture (SASA), funded by the European Social Innovation Funds and the Scottish Government under project number SIF-R5-S2-HI-004.

S. Blackley (✉) · D. McVey
Queen Margaret University, Edinburgh, UK
e-mail: sblackley@qmu.ac.uk

M. Scholten
Scottish Crofting Federation, Kyle of Lochalsh, UK

A. Veitch
Doughies Bakery, Fort William, UK

Y. Nishikawa and M. Pimbert (eds.), *Seeds for Diversity and Inclusion*,
https://doi.org/10.1007/978-3-030-89405-4_9

livestock feed. This multi-author study presents and analyses findings into the crop's potential as the raw material for locally produced flour, bread and beer, offering new opportunities in sustainable seed saving, small-scale agriculture, food production and eco-enterprise. The authors—part of the project's multidisciplinary team of researchers, artisanal food producers and crofters—explicate aspects of the pioneering project, from conditions on Uist's coastal machair where the rye originates, to testing seasonal varieties in mainland Lochaber and assessing nutritional qualities and consumer acceptance of novel products. They conclude that Hebridean rye, with its potential for crofters in remote locales and local businesses, could help in preserving agrobiodiversity, traditional knolwedge and practices, crofting culture and economic resilience in the north and north-west of Scotland.

Keywords Artisanal products · Collaborative research · Economic resilience · Rye landraces · Scottish crofters

9.1 INTRODUCTION

This chapter describes a project exploring the potential for adding value to the landrace, Hebridean rye (*Secale cereale*) by finding new culinary and food-related uses for it, with a view to reintroducing it to the mainland Highlands of Scotland for cultivation. Hebridean rye was traditionally grown in Scotland's Highlands and Islands—the mainland Highlands, along with Orkney, Shetland and the Outer Hebrides—but is currently restricted to the Outer Hebridean islands.

Hebridean rye is a component of a livestock feed mixture valued for its local adaptation and good yields under extreme conditions. It is cultivated by crofters, who work small landholdings, or crofts, that are unique to Scotland, and deploy low-input practices. Cereal cultivation on islands can be viewed as a unique form of agricultural biodiversity, connecting traditional and cultural practices with place-specific land management processes. As cultivation of the rye has reduced in the Highlands, the traditional knowledge and agricultural systems linked to it are being lost.

This multipartite and interdisciplinary research was therefore significant. Moreover, a team of researchers and artisanal food producers was put in place to assess the landrace's potential for novel uses. Inspired by seed and food sovereignty values, the project sought to explore new opportunities for and with crofters to cultivate the rye for use in new food and drink products such as flour, bread and beer. These types of opportunities are vital to the preservation of crofting life, and to support young crofters in turning the demographic tide in the Outer Hebrides where the population is ageing.

Uist in the Outer Hebrides, a group of six islands, is one of the few areas in Scotland and the United Kingdom with surviving landraces. There are written records for their production dating back to the seventeenth century (Martin, 1703). To this day, a mix of barley, oats and rye are still grown as 'crofters' corn', used as animal feed.

The endurance of this traditional cereal mixture may in part be due to the unique conditions of the <u>machair</u>—the unique coastal flatlands habitat where the crop is grown. Highly alkaline soils block the uptake of essential nutrients, such as manganese, by crops, leaving them prone to disease and stunted growth and there are currently no commercial cereal varieties on the market in Britain that can cope with these extreme conditions without additional spraying with manganese, yet the local landraces are able to yield without this intervention due to local adaptation (Schmidt et al., 2019). The machair is also protected by a number of environmental designations and restrictions, which can disallow the requirement to apply manganese spray on 'non-island varieties'. This combination of factors makes the native landraces the most economical to cultivate. The Hebridean rye landrace is therefore a 'crofters' landrace'.

9.2   Seed Sources and Seed Governance in the Highlands and Islands

With the islands of Uist entirely self-reliant on locally produced seed in a low-input agricultural system, seed governance here can be seen as an informal seed system, or perhaps more accurately, ad hoc seed supply chains without formal organisation or central point, in other words, a 'spontaneous order' (Ward, 1988) based on values of crofter-to-crofter mutual aid rather than for-profit seed production. The seed itself is traded as it is grown: as a mixture, pivotal to keeping the entire livestock rearing system feasible. The seed, as the source of home-grown fodder, reduces the cost of imported silage, as observed by crofter Angus Laing in *It's In the Blood*, a 2018 BBC documentary on crofting on South Uist. Furthermore, there is a regional ethos based on seed saving as community service in the face of economic necessity. This ethos can be seen as a form of mutualism to support the precarious island crofting life, sometimes still through barter, in the knowledge that this is a unique seed system. Part of the ethos is locals' pride in producing 'a good clean seed crop', similar to that observed among the Northern Isles crofters on Orkney and Shetland in Mahon's study (2016).

In recent decades seed growing, once a feature of every crofting township, has become more precarious. Fewer crofters are involved due to a combination of increased pressure on seed supply by pest species such as geese and deer, fewer hands (beyond demographic shifts, there is a lack of local employment and housing to retain younger locals), declining knowledge and experience and a narrower harvesting window due to the impacts of climate change.

Research on Orkneybere barley has contributed significantly to its preservation on Orkney, but prior to the present project, no similar work had been done on Hebridean rye. The local threats make research on the Uist landraces more urgent, as opening up new value chains for landraces can contribute to their preservation as well as create crucial income opportunities. The latter are vital not just for young crofters but also for entrepreneurs such as the team behind the North Uist Distillery, who are keen to build on island crofting traditions. The distillery's founder and artistic director Kate MacDonald has written on their website that they are 'proud to be able to promote a new generation of crofting—providing people with a reason to grow the island's trademark organic bere barley', the most successful Scottish landrace (*Hordeum vulgare*). These emerging, often micro- or social enterprises, are seen as vital to turning the demographic tide on Uist (Fisher & Morrison, 2019).

Adding value to traditional varieties has succeeded on Orkney both with bere, and with traditional Scottish wheats (Whitley, 2019), through the development of value chains for flour and beverages, and assessments of the flours' nutritional profile (Theobald et al., 2006). The ongoing work on bere on Orkney has become a model for other areas, new distilleries on the Inner Hebridean island of Raasay, for instance, and has inspired the present Uist project.

Scotland's traditional cuisine does not feature rye bread, despite the popularity of rye in nearby Nordic nations. Although commonly grown at scale in other countries such as Germany, Poland and Estonia, it is currently only grown on a small scale in Scotland. The name 'Hebridean rye' requires some explanation. When rye samples were collected in 2009 as part of gene-bank conservation of Scottish landraces (Scholten et al., 2009), the samples were deposited at Science and Advice for Scottish

Agriculture (SASA) in Edinburgh under the name Hebridean rye. From that time, the Uist rye landrace was referred to thus, in addition to its vernacular Gaelic name, *seagal*. Scottish Gaelic is still the dominant language over much of the Outer Hebrides and remains common as the working language of crofters.

The Gaelic names of the other two cereal landraces passed down and still in use are for oats (*Avena strigosa*) either *coerce* or *coirce beag*, while for bere (*Hordeum vulgare*) is *eorna*. Both names have an inherent distinction from seeds from the mainland. 'Coirce mor' and 'mainland barley' for other barley varieties. A recent example of the vernacular use of landrace names can be heard in the documentary *Crofting and the Uist Machair* (Farming Advisory Services, 2019), in which North Uist crofter Donald John MacDonald notes: 'We sow what we call *coirce beag* and sometimes some rye'.

9.3   Lochaber's Demand for New Rye Landraces

Rye is currently growing in both recognition and popularity in Scotland. An increasing number of small and artisanal bakers are interested in traditional and landrace varieties of cereal that may not necessarily suit large-scale processing, but provide the opportunity for a more flexible and innovative approach to the use of flour mixes and bakery products. In recent years, craft brewing and distilling have seen significant growth in Scotland, with small-scale producers satisfying increasing public demand for diversity within the market, and rural and island breweries and distilleries seeking new ingredients that help them to connect with their local environment and heritage, or that have an interesting story to tell and sell.

One of the crofters involved in the project, Adam Veitch, is developing his existing commercial micro bakery into a 'seed to loaf' peasant model of bakery with rye grown on local crofts within Lochaber, near Fort William in the Highlands (Veitch & Veitch, undated) A rye trial was proposed to explore the potential of different varieties and in particular Hebridean rye for growing in the Lochaber area on the Scottish mainland, and to assess its potential in rye breads and other baked products.

There is a burgeoning interest in local food in Lochaber, and a new ethos in supporting the re-emergence of a local food economy. It is partly driven by tourism, but also by local growers and consumers, and organisations such as the Lochaber Environmental Group and Food Lochaber, a collective of local producers that are devoted to production informed by organic principles and crofting practice. Lochaber is not a locale associated with farming traditions. Grain growing has largely disappeared over the last century deforestation has led to soil degradation (Wombell, 2003), predatory deer numbers are rising and a lack of grain-specific machinery and know-how have made reintroducing cereal growing very challenging.

However, the factors detailed in this chapter—the survival of an unexplored rye landrace on Uist, and the search for new rye landraces on the mainland—inspired the Crofters' Diversity Pays! (CDP) project. This initially lasted a year (May 2019–April 2020), but remains ongoing to date because of the covid-19 pandemic.

9.4   Project Aims and Research Questions

Under the CDP project, researchers have examined potential income streams for crofters through the assessment of new uses for traditional landrace varieties they grow, many maintained over generations as a unique form of agricultural biodiversity. Developing new uses and high-end products from low-value agricultural assets can provide new business opportunities for crofters and develop new markets for their produce which, in turn, can help both maintain and reinvigorate a traditional way of life.

The project aimed to undertake research in a crofting context, with crofters involved in the investigation, and also as beneficiaries, by assessing the potential value of a neglected landrace for a potentially new growing area.

The aim of the rye workstream was to investigate the potential to add value to Hebridean rye in order to inform crofters and food and drink producers of the potential uses of the cereal as a new and unique ingredient in a number of artisan products. The research questions included the following:

- How does Hebridean rye compare with other varieties of rye as regards its suitability for a mainland grain-to-loaf value chain
- What kind of processing can be used to transform Hebridean rye into food and drink products?
- How might the introduction of Hebridean rye be leveraged to create a local grain supply for food and drink production?

- Does the rye have special nutritional value, does it taste good and will people eat it?
- What end-use food and drink products might be produced using the rye?
- How might these Hebridean rye products be branded and marketed?

9.4.1 The Chosen Locations

In the course of the project, the Hebridean rye seed grown in Uist was introduced in <u>Lochaber</u>. As a consequence, issues of continuity, and discontinuity, of tradition and values related to seed arose between the two locations.

Uist and Lochaber can be seen as representing two approaches to contemporary crofting practice. On the one hand is the long-established Outer Hebridean traditional crofting mode of agriculture, an island practice based on a livestock rearing system; local seed sources are intact, and it is run by mostly primary producers selling to mainland markets, in a linear model with very long chains. On the other hand is a mainland Highlands practice emerging where cereal growing and local seed sources have disappeared, and where groups of collaborators organise direct sales for crops and meat using agroecological practices and ethos. The general attitude on the mainland is that reintroduction of locally grown grain will be both challenging but ultimately beneficial.

In both locations, producers sought collaborations with local processors who might be interested in developing new products from Hebridean rye. Results of the research were also published in local newspapers in both locations.

9.4.2 The Growing Trial in Lochaber

A number of rye varieties were sourced as comparison varieties for benchmarking against the Hebridean rye and also with a view to building a rye nursery in Lochaber with a wider range of varieties. The ultimate goal was to create a locally adapted and evolving rye population that would yield enough grain for a local micro bakery as well as provide a model for other growers and micro producers In addition, participants running the rye nursery formed observation learning material for the crofters to begin to communicate and restore the know-how of growing grain in Lochaber.

Both summer and winter varieties were sourced and tested, despite there being no recorded history of growing winter cereals in this part of the Highlands. This may have been due to the arable fields being used for winter grazing.

During the 2019 summer growing season at the croft nursery, three different rye varieties were grown: Hebridean rye, a commercial variety and a landrace originating from Scandinavia. During the 2019/2020 winter, another five varieties were tested: one Baltic, three Scandinavian, a mixed population sourced through the collaborative project Scotland the Bread and a traditional German bread rye.

The Hebridean rye was found to be suitable for growing in Lochaber, and to have some useful attributes as a crop that may work well in developing an evolutionary cereal rye population for the area (Veitch & Scholten, 2020). However, overall, the winter rye varieties outperformed the summer varieties on both weed suppression and yield.

This growing trial was an important strand of the overall project, exploring the pathway from seed to harvest and providing practical insight from the perspective of the crofter that grew the crop. And, beyond the more specific narrow aim, it was key in building genetic resources, the knowledge and experience of growing rye, and an acquaintance with different rye varieties and their suitability for Lochaber growing conditions.

9.4.3   Sensory Qualities, Nutritional Value and Product Testing

The Hebridean rye was tested to ensure that levels of ergot and other pathogens were safe, and that it was of high quality. Artisanal bakers were recruited to test its milling properties and viability for baking; craft brewers explored its suitability for brewing. All results were positive, confirming that the rye is suited to small-scale flour production, artisan baking and the production of craft alcohol products.

Other research has suggested that traditional or landrace crops can have higher nutritional value than modern cultivars. This was the first research to report the nutritional properties of Hebridean rye, revealing that it contained high levels of many useful micro- and macro-nutrients and had potential for use in healthy food products, and as an added ingredient to fortify other foods. The rye was found to be high in fibre and protein; it also contained significant levels of phosphorus, manganese and zinc, and has a diverse mineral composition overall.

A number of prototype and test breads and beers were produced at the Lochaber-based 'Doughies' and 'Grain of Truth' bakeries. Initial results from a series of consumer tasting panels undertaken through the MSc Gastronomy department at Edinburgh's Queen Margaret University suggested that the products had a distinctive and enjoyable flavour profile and that consumer acceptance levels were high, suggesting that people would buy products containing Hebridean rye.

Branding and packaging inevitably influence consumer choices. Initial discussions suggested that there was significant potential for marketing products containing Hebridean rye, and that the stories behind its nutritional qualities, scale of production, crofting heritage, Gaelic connections and specific provenance and terroir would be key to the process.

9.5   Conclusions

This was the first investigation into the qualities of the rye landrace from Uist, and potential products derived from it. The research had an artisan crofting-baking business as starting point and developed into a partnership of academic researchers, food processors, small-scale food and drink producers, and established new links between crofters across remote areas in the Highlands and Islands of Scotland. The research has been vital to understanding the journey towards marketable products based on Hebridean rye, which, if developed, could ensure its viability as a crofting crop along with that of other traditional landraces.

One of the first and unintended outcomes of the research was that from the very start it showed how little known the existence of a local Scottish rye was, even in its Outer Hebridean place of origin. Outside that region, few growers and processors were aware of it, although Scotland has recently seen strong interest in rye as an ingredient for baked goods and alcoholic drinks. This confirmed the importance of research into rye as an underutilised crop and, more generally, the need to foster more awareness of local Scottish varieties of rye.

Secondly, the growing trial looked into the potential of Hebridean and other rye landraces as a crop for agroecological crofting in a new environment and as a new seed source. Trials such as these—which take place outside agricultural institutions, focus on minor crops and are led by farmers working at a small scale in marginal growing areas—need further support. Such approaches are more common on the European continent. The similar but much larger Diversifood project, and agricultural biodiversity networks such as Let's Liberate Diversity,[1] were consulted as models for this project.

Regarding seed governance and ethos, there seemed to be much in common between the Hebridean informal ad hoc seed supply based on a mutual support ethos and the new location in Lochaber. As Adam Veitch noted on plans for his growing programme, the rye 'needs to go on a bulking up exercises the next few seasons, both for ourselves and to share/distribute to others. I'd love this to be open source but it also needs to be sustainable'. His plan is to create a rye population hefted to an area, evolving and becoming its own locally adapted grain, rather than strict adherence to the Hebridean rye cultivar.

This groundbreaking research project has provided information that can be utilised to provide new business opportunities for crofters and to develop new markets for their produce. It started investigating each step in the value chain of Hebridean rye from croft-grown rye as input for a micro-artisan bakery and several beverages. The research uncovered various solutions to the lack of machinery and equipment needed to sort, clean, and dry seeds. Based on their own agricultural needs, crofters innovate by constructing their own machinery using engineering skills and traditional knowledge. Access to commercial or laboratory seed cleaning machinery is limited and costly. The research identified the need for machinery infrastructure at each step in the value chain, both in its original and new location.

Hebridean rye has the potential to play a valuable role in Scotland's food and drink industry, providing business opportunities for those living in remote and rural areas, allowing the diversification of supply chains and potentially even impacting positively on dietary health. An added benefit is the positive impact that this could have on the preservation of agricultural biodiversity and the traditional knowledge and practices associated with the cultivation of Hebridean rye.

Acknowledgements Crofters on Uist are warmly thanked for their time, seeds and advice. Andrew Whitley (Scotland the Bread) for advice, seeds and data. Anders Naess of Specialkorn in Norway, Annika Michaelson of Mustiala Agricultural School in Finland, and the Estonian Crop Research Institute for making seeds available.

---

[1] www.liberatediversity.org

References

Diversifood project. www.diversifood.eu. Accessed July 25, 2021.

Farming Advisory Services. (2019). *Crofting and the Uist Machair*. Video documentary. Retrieved December 15, 2020, from https://www.youtube.com/watch?v=m7MP0D2YyDU

Fisher, T., & Morrison, T. (2019). Turning the tide on Uist. Webblog *Islands Revival*. CoDeL. Retrieved July 25, 2021, from https://islandsrevival.org/turning-the-tide-on-uist/

Let's Liberate Diversity. https://liberatediversity.org/ as Accessed July 25, 2021.

Mahon, N., McGuire, S., & Islam, M. M. (2016). Why bother with *bere*? An investigation into the drivers behind the cultivation of a landrace barley. *Journal of Rural Studies, 45*, 54–65. https://doi.org/10.1016/j.jrurstud.2016.02.017

Martin, M. (1703). *A description of the Western Isles of Scotland circa 1695* (1999 ed.). Birlinn Press.

North Uist Distillery Co. website (undated). https://www.northuistdistillery.com/. Accessed July 25, 2021.

Schmidt, S. B., George, T. S., Brown, L. K., Booth, A., Wishart, J., Hedley, P. E., Martin, P., Russell, J., & Husted, S. (2019). Ancient barley landraces adapted to marginal soils demonstrate exceptional tolerance to manganese limitation. *Annals of Botany, 123*, 831–843. https://doi.org/10.1093/aob/mcy215

Scholten, M., Spoor, C., & Green, N. (2009). Machair corn: Management and conservation of a historical machair component. *The Glasgow Naturalist, 25*, 63–71. Supplement: machair conservation: successes and challenges.

Theobald, H. E., Wishart, J. E., Martin, P. J., Buttriss, J. L., & French, J. H. (2006). The nutritional properties of flours derived from Orkney grown bere barley (*Hordeum vulgare* L.). *Nutrition Bulletin, 31*(1), 8–14. https://doi.org/10.1111/j.1467-3010.2006.00528

UHI.   https://www.uhi.ac.uk/en/research-enterprise/res-themes/silk/agronomy-institute/research-on-cereals/

Veitch, A., & Scholten, M. (2020). Rye nursery technical report. Unpublished report as part of final report of the CDP! Project to the Scottish government.

Veitch, A., & Veitch, A. (undated). Doughies. Retrieved July 25, 2021, from www.doughies.blog

Ward, C. (1988). *Anarchy in action*. Freedom Press. (Reprint from 1973 first edition).

Whitley, A. (2019). *The nutritional profiles of heritage wheat and rye varieties grown by Scotland The Bread in 2018*. Scotland The Bread. Unpublished.

Wombell, J. (2003). *The spirit of the soil: An historical perspective on soil fertility and the application of composting in Lochaber*. Lochaber Environmental Group.

Wood, B. (2018). *San Fhuil/It's in the Blood*. Documentary. TrixPixMedia. Retrieved July 24, 2021, from BBC ALBA website. https://www.bbc.co.uk/programmes/m0001bdf

Inside the Japanese Seed Industry: Its Characteristics and Implications for Agroecology

Ayako Kawai

Abstract The nature of small-scale seed companies and their role in sustaining genetic diversity are understudied in developed countries—not least Japan, which has nearly 1000 of them. In this in-depth survey, Ayako Kawai analyses findings derived from interviews with the heads of, and breeders in, three such firms. Historically, she notes, Japanese seed companies operate within close-knit networks and follow customary practices, which has helped to foster cooperation: they function as "diverse economies", collectively agreeing on seed prices and trading as equal partners. Many of them also contribute to agrobiodiversity by maintaining open pollinated varieties, and in some rare cases, by developing new non-hybrid ones. Inevitably, these practices can put them at odds with market constraints such as the demand for mainstream traits. Kawai concludes that if free-market principles were applied wholesale to Japan's seed industry, its predominantly value-led approach would suffer, with negative impacts on national crop diversity.

A. Kawai (✉)
Research Institute for Humanity and Nature, Kyoto, Japan

© The Author(s) 2022
Y. Nishikawa and M. Pimbert (eds.), *Seeds for Diversity and Inclusion*,
https://doi.org/10.1007/978-3-030-89405-4_10

Keywords Agroecology · Breeding · Customary practices · Fair Trade Commission · Seed companies

10.1   INTRODUCTION

A stable, high-quality, locally appropriate seed supply is a key element of sustainable agri-food systems. Yet recent decades have seen a massive consolidation of seed companies into huge multinational agribusinesses, sparking concern over their increasing control of the food chain as a whole, from seeds to market shelves (Howard, 2009). Over the last half-century, this development has transformed the sector: once composed primarily of small-scale family businesses, it now features a limited number of multinational pharmaceutical and chemical companies (Fernandez-Cornejo & Just, 2007; Vellve, 2009).

Without significant changes in the global political environment, the trend for consolidation is expected to continue (Howard, 2009). Such a development may further exacerbate the difficulties of instating renewable agriculture by reducing choices among farmers to obtain seeds that are locally adapted, genetically diverse, novel, non-patented and compatible with self-reproduction (Howard, 2009; Schimmelpfennig et al., 2004). In addition, the market clout of multinational pharmaceutical and chemical companies could exclude seed suppliers, who are not bound to narrow economic goals and are committed to providing diverse seeds.

Studies suggest that a number of emerging seed companies offer heirloom or open pollinated seeds in response to increasing interest among gardeners and organic farmers in the United States (Bonina & Cantliffe, 2004; Nabhan, 2013). However, the characteristics of local small-scale seed companies and their role in maintaining and distributing genetic diversity are understudied, especially in developed countries.

This chapter explores the nature of the Japanese seed industry and describes the motivations of, and decisions made by, the heads of small-scale family-owned seed companies. It also examines their contributions to agroecology, if any. It is based on semi-structured interviews undertaken in 2016 and 2018 with the heads and breeders of three such companies. Since some of these did not want to be identified, their names, location and the type of crops they breed, are kept confidential, barring Noguchi Seed for its name.

10.2 Characteristics
of the Japanese Seed Industry

The global seed industry is polarized. On the one hand are large-scale integrated agribusinesses; on the other, as with most Japanese seed companies, are independent operations. In Japan, such companies maintained independence from the huge multinational agribusinesses by remaining privately owned and avoiding any listing on the stock market. One of the interviewees noted that "it is Japanese culture to refuse selling their companies to foreigners even if it leads to bankruptcy". The modest size of Japanese vegetable seed markets also made them less attractive to foreign investment.

Intensified market competition and advancement in breeding technology have resulted in the rise of large seed companies and the closure of smaller ones in Japan (Hisano, 1998). Yet there are nearly 1000 seed companies registered in the country, including retailers (JSTA, 2020)—a high number compared to that in other developed nations.[1] With the production and development of major crop seeds such as rice, wheat and soybean, regulated under the Main Crop Seeds Act until 2018, most Japanese seed companies have engaged in vegetable breeding (Hisano, 1998; Matsuura, 2012). The exact number of such companies is uncertain, but it seems that nearly 40 breed Brassica varieties, for instance.

It is the Japanese seed industry's customary governance structure that seems to support the coexistence of diverse domestic vegetable breeding companies. Before seed companies were established in Japan, specialized farmers with high-level agricultural skills bred new varieties (Abe, 2015). Such producers became seed dealers in the late seventeenth century, and by the end of the nineteenth century had established the first modern seed companies (Abe, 2015). Since each company specialized in breeding one crop, they sold seeds to each other to ensure that customers enjoyed a wide range of varieties. Customary business rules, including selling seeds to each other, had been established around the late eighteenth century (Abe, 2015), and are still in place. Thus, breeding companies not only sell seeds directly to customers; they also operate as a wholesaler and sell seeds to other companies. When seeds are sold through wholesalers

---

[1] Nabhan (2013) reported that there are at least 275 vegetable seed companies in the United States and Canada.

and retailers, customers do not know which company originally bred the variety, since its name is determined by the end sellers.

By specializing in a certain kind of crop, vegetable breeding companies have collected and developed breeds over time, which has been a distinct benefit for old, established seed companies. The head of Yamamoto, for instance, commented that having good breeding lines makes a company competitive. A breeder at Yamamoto mentioned that there are probably fewer than a dozen breeders in Japan who deal with the same crop he does. While a relatively small business, Yamamoto is competitive in terms of crops they specialize in; the world's seventh-largest seed company, Sakata in Japan, does not specialize in these crops and has become their trading partner. Large seed companies such as Sakata, which occupy much of the domestic vegetable seed market, are also large wholesalers and customers of smaller breeding companies.

Seed companies, including large ones, follow customary practices in their operations. A representative of one explained that since the Meiji era in Japan (1868–1912), major seed companies across the nation have organized regular gatherings to trade seeds through bidding, deciding sales prices, and exchanging information about market trends and the quality and amount of seed production. The Japan Fair Trade Comission regards this as cartel pricing—that is, agreed pricing by a group of producers working together to protect their interests, which restricts free-market competition.

While I did not interview anyone in detail regarding this issue, existing studies and rulings by Japan's supreme court suggest that the 32 seed companies—which provide more than 90% of domestically distributed seeds of four Brassica varieties—did engage in cartel pricing by agreeing a basic seed price at least between 1998 and 2001 (Wada, 2009). These companies were members of a specialized Japan Seed Trade Association subcommittee made up of the country's breeding and seed producing companies.

During an annual subcommittee meeting, the companies voted to determine the "standard retail price" of four crops. This became the basis for setting other standard prices, for wholesale and for selling to agricultural cooperatives, for instance. Members voted on whether to increase, decrease or keep the standard retail price from the previous year, and then decided the actual standard retail price. The companies then adjusted their seed prices using the ratio of changes in the standard retail price from the previous year. Discounts and other adjustments were sometimes made for individual trading partners, depending on the history of the transaction, volume of trade and transaction value (Wada, 2009). One interviewee mentioned that the annual gatherings of seed companies were slated to end later that year of 2018. It is unknown at this stage how such a change might influence this customary practice.

Links between seed companies were not limited to business relations but sometimes extended to familial relations. The head of Hakonishi explained that marriages between members of the families of seed companies were common, for instance. One of the heads interviewed noted that the industry is "close-knit" and operates like "a single family", adding: "Seed companies have a long-term relationship across two or three generations, so we are closer than relatives". He speculated that such relationships between seed companies could be unique to Japan.

The way Japanese seed companies operate, within close networks and based on customary practices, might be a factor in enabling the coexistence of various domestic companies and regulated business mergers by foreign companies.

10.3   Small Seed Companies and Agroecology in Japan

With the rise of breeding technology and policy reforms for mass vegetable production in Japan, vegetable breeding has shifted from developing open pollinated to hybrid varieties (Abe, 2015). Many local seed companies, facing business closures or lack of demand, no longer maintain local varieties. That could potentially lead to the loss of crop diversity.

The seed company heads interviewed noted that they face a dilemma between strictly defined market demands and their preferences to maintain old varieties or develop unique ones. The head of Yamamoto mentioned that he could no longer breed varieties with traits that he liked due to changes in market demand. Even though he valued the sweetness and soft texture of a certain crop, other market requirements need to be met, such as transportability, sturdiness and suitability for retail packaging. He noted that "it is painful that even though there are much tastier crops", the market effectively prevents companies from breeding them.

Interviewees expressed a strong desire to breed new varieties but noted that the seed companies' attitude towards pursuing uniqueness was a source of tension. Breeding new varieties was perceived as a risk in the face of conservative consumer preferences and inflexible market standards.

Since consumers tended not to be adventurous in their choice of vegetables, the characteristics of popular crops did not change much. Breeding companies perceived it as a risk to develop varieties with new features that consumers were unfamiliar with. The head of Yamamoto confirmed the power of the market and consumer preferences—rather than farmers or seed companies—in determining the characteristics of produce. This led to seed companies, including Yamamoto, feeling that their autonomy in decisions over which varieties to maintain and develop was limited.

Despite these hurdles to maintaining and breeding niche varieties, interviewees from small seed companies developed and sold local, open pollinated and unique varieties that do not reflect mainstream market trends. Two heads mentioned farmers' needs for old varieties as the main reason to maintain them even when they do not make a profit. The head of Yamamoto explained that since "seed companies do not operate for sake of genetic resource conservation", they dropped several varieties from their catalogues every year. However, he stated that they maintained some old varieties even if the company did not profit from them, noting: "I feel that I need to provide seeds that our customers request every year". Similarly, the head of Hakonishi said that the company maintained local open pollinated varieties farmers asked for. He saw it as a duty since domestic varieties are culturally important and "it is also our task to maintain them".

Apart from the needs of farmers, and a sense of duty to preserve old varieties, personal and familial reasons also influenced seed companies' decisions to maintain local open pollinated varieties. The heads of both Hakonishi and Yamamoto maintained varieties that were developed by their fathers or grandfathers. When I asked the head of Yamamoto whether he, like traditional farmers, maintained a variety because it was handed down from their ancestors, he said yes. Some seed companies operate based on a traditional family system, similar to that of traditional farmers (see Chapter 5). Under this system, the continuation of *ie.*, or the household, is the top priority. Even private companies have adopted this approach, which emphasizes to sustain the household and their businesses and prefers taking a long-term perspective rather than maximizing profit in the short term (Teranishi, 2018). Since the seed companies researched operate as family businesses, non-monetary norms and values played an important role in their decision-making.

I found that the seed companies studied not only maintained old open pollinated varieties but also developed novel non-hybrid and minor varieties. Unusually for a seed company, Hakonishi occasionally released new open pollinated vegetable varieties developed from genetic materials that did not suit hybrid breeding. Yamamoto also actively developed unique varieties because "it is our pleasure to introduce" them. While breeding for mainstream traits was crucial, it was "not fun", so they also bred for minor or unique varieties. Yamamoto also bred a variety specifically suited to the environment and culture of the Hokkaido region. The company involved local farmers in the breeding process and sought their advice when choosing preferable lines.

Seed company interviewed also accommodated agroecological value in their business practices. Noguchi Seed sells only open pollinated seeds. Besides developing, maintaining and selling a local open pollinated turnip variety, they also sold open pollinated seeds from other companies. Given that customers can find it difficult to distinguish between hybrid and open pollinated varieties, Noguchi was popular among seed savers. The company also sold seeds online as well as in retail outlets, enabling wide distribution of varieties that are otherwise difficult to find. Its head, who emphasized the importance of seed saving for the conservation of agrobiodiversity, has also written books on how to save purchased seeds.

Breeding companies currently outsource the production of up to 95% of seeds to other countries (Matsuura, 2012). Hakonishi, which outsources 80% of its seed production outside Japan, has decided to increase domestic seed supply by up to 50% by establishing a new seed producing company. Their decision is a response to production failures triggered by unexpected extreme weather events. While such changes would increase production costs, the head of Hakonishi noted that "it is important for farmers to secure seeds that they want, and it is our duty to provide a stable seed supply". This project is run under the government funding scheme for adding value to agricultural production and Hakonishi funded 50% of it to establish its new seed producing company.

The head of Yamamoto expressed frustration over how seed companies' public image reflects that of multinational pharmaceutical and chemical companies. He noted that such multinationals "integrate chemical products and seeds", a business model that is "definitely different from ours", adding: "People think that we are the bad guys".

10.4   Discussion and Conclusion

The commodification of seed in the global food regime has concentrated power among limited number of multinational agribusinesses. Members of the food sovereignty movement see the free trade philosophy as destructive to local economies and call for a redefinition of mechanisms for market, trade and exchange to enable producers and consumers to retrieve democratic power over food production and distribution (Nyeleni Movement for Food Sovereignty, 2007). The small-scale seed companies featured in this chapter maintained, developed and distributed niche varieties, including old local non-hybrid varieties, with an eye on values other than profits. Thus they demonstrated their potential engagement in an agroecological seed system that could benefit farmer resilience and agrobiodiversity conservation.

The members and heads of small seed companies who were interviewed maintained old varieties and developed unique varieties despite their lack of value in the official market. For those companies, seeds were not reduced to commodities but linked to family history and a sense of duty to offer high-quality seeds for the public good. Collectively, they show their commitment to communities by voluntarily producing and stocking vegetable seeds in case of catastrophe—a task previously funded by the Japanese government.

Unique customary practices among seed companies, especially mutual trading, have allowed them to provide diverse varieties to farmers. And by each specializing in different seeds, they have enabled the coexistence of many companies. Under this arrangement, companies traded their goods as equal partners regardless of their size. While further investigation is needed, customary practice to collectively decide a "standard price" might have favoured small breeding companies with limited resources. The trading system used by seed companies enhanced farmers' access to diverse types of seeds, as they could not always source local varieties within their communities (see Chapter 5). By selling seeds that can be bred from, and by not imposing patent rights on developed varieties, seed companies have not restricted farmers' access to genetic diversity.

Some heads expressed frustration over neoliberal market constraints and norms, which have eroded their autonomy in making decisions on crop characteristics. Even when seed companies are highly motivated to conserve old varieties, the lack of market demand has been demoralizing. For now, old varieties are maintained to some extent by small seed companies, motivated by a sense of commitment to traditional values. Yet varieties could still be disappearing, given the harshness of the business environment. As the domestic market shrinks along with a diminishing population, seed companies will need to expand their engagement with the global seed economy.

Seed companies have operated as a loosely tied social group regulating how and to what ends seeds are distributed; competition over prices has been limited. This has been regarded as problematic by the Japan Fair Trade Commission for promoting free-market capitalism. Yet another way to interpret the way seed companies operate, including through mutual agreement on seed price, is that they function as "diverse economies" (Gibson-Graham et al., 2013). By collectively deciding on seed price, they have possibly ensured solidarity among breeding companies, including small ones, and diversified options for farmers. That begs the question of how such a system could be sustainable under a strict neoliberal policy environment.

Meanwhile, it is worth noting that the Japanese seed industry was under direct government control during the Second World War; there was a merger of seed companies, and seed producers and prices were determined by the government. That has inevitably influenced the features of the informal governance structure that currently exists within the seed industry (Okada, 2005). It is also important to note the recent trend among larger seed companies to breed infertile hybrid varieties, which do not produce pollen, to prevent other companies from using them for breeding. That has sparked concerns about genetic "enclosure" within such companies.

The aim of this chapter is not to romanticize seed companies, nor to reduce concerns over the intensifying control of genetic resources among the more capital-intensive companies. However, existing studies have overlooked the other end of the spectrum, especially small, locally based companies that contribute to sustainable agri-food systems. If free-market economic principles were strictly applied to the Japanese seed industry, within which small seed companies play a crucial role, the process could dismantle the informally established governance structure. Such an intervention would damage the coexistence of diverse seed companies and the distribution of diverse varieties across the nation, which would diminish Japan's crop biodiversity.

Acknowledgements The research has been supported partially by Research Institute for Humanity and Nature (RIHN: A constituent member of NIHU), FEAST Project (No. 14200116) and JSPS KAKENHI 17H04627.

References

Abe, N. (2015). *Dentōyasai wo tsukutta hitobito* [People who developed traditional vegetables: A modern history of 'seed dealers']. Nōbunkyō.

Bonina, J., & Cantliffe, D. J. (2004). *Seed production and seed sources of organic vegetables.* Horticultural Sciences Department, Florida Cooperative Extension Service, IFAS. University of Florida. https://doi.org/10.32473/edis-hs227-2004

Fernandez-Cornejo, J., & Just, R. E. (2007). Researchability of modern agricultural input markets and growing concentration. *American Journal of Agricultural Economics, 89*(5), 1269–1275. https://doi.org/10.1111/j.1467-8276.2007.01095.x

Gibson-Graham, J. K., Cameron, J., & Healy, S. (2013). *Take back the economy: An ethical guide for transforming our communities.* University of Minnesota Press.

Hisano, S. (1998). Shubyōjigyō no kōzō to kinō ni kansuru ichikōsatsu [The structure and functions of the Japanese seed system]. *Review of Agricultural Economics, 54,* 21–37.

Howard, P. H. (2009). Visualizing consolidation in the global seed industry: 1996–2008. *Sustainability, 1,* 1266–1287. https://doi.org/10.3390/su1041266

JSTA. (2020). About the association. Retrieved February 19, 2021, from http://www.jasta.or.jp/about/

Matsuura, M. (2012). *Shushi sangyō: ninaite no henka to sijyō no kakudai* [Seed industry: Changing stakeholders and market expansion]. Mitsui & C. Global Strategic Studies Institute. Retrieved July 24, 2021, from https://www.mitsui.com/mgssi/ja/report/detail/__icsFiles/afieldfile/2016/10/20/120720i_matsuura.pdf

Nabhan, G. P. (Ed.). (2013). *Conservation you can taste: Best practices in heritage food recovery and successes in restoring agricultural biodiversity over the last quarter century.* University of Arizona Southwest Center/Slow Food USA. Retrieved July 24, 2021, from https://garynabhan.com/pbf-pdf/ConservationYouCanTaste.pdf

Nyeleni Movement for Food Sovereignty. (2007). *Nyéléni 2007: Forum for food sovereignty.* Retrieved January 6, 2022, from https://nyeleni.org/DOWNLOADS/Nyelni_EN.pdf

Okada, T. (2005). Senjika no nōgyō shizai mondai [The agricultural materials problem during wartime]. *Agricultural History Society of Japan, 39*, 11–22. https://doi.org/10.18966/joah.39.0_11

Schimmelpfennig, D., Pray, C. E., & Brennan, M. (2004). The impact of seed industry concentration on innovation: A study of US biotech market leaders. *Agricultural Economics, 30*(2), 157–167. https://doi.org/10.1111/j.1574-0862.2004.tb00184.x

Teranishi, J. (2018). *Nihongata shihon shugi* [Japanese style capitalism]. Chūō-Kōron shinsha.

Vellve, R. (2009). *Saving the seed: Genetic diversity and European agriculture.* Earthscan.

Wada, T. (2009). Shushi kakaku karuteru shinketsu torikeshi soshō jiken hanketsu no kentō [Evaluating the judgment in the case of the revocation of the trial decision on the seed price cartel]. *NBL Antitrust Law Case Study Group, 914*, 63–70.

CHAPTER 11

Bhutan's 'Middle Way': Diversification, Mainstreaming, Commodification and Impacts in the Context of Food Security

Mai Kobayashi

Abstract The Himalayan kingdom of Bhutan is both wedded to tradition and influenced by the global push to modernize. In this study of the country's path to food security, Mai Kobayashi describes its evolving national 'middle way' towards sustainable agriculture. She traces seed-sector dynamics over the past 70 years, as exogenous influences from India and Japan mingled with endogenous practices. First following a Green Revolution-style high-input agricultural model reflecting India's, Bhutan joined the Colombo Plan in 1962, paving the way to autonomous economic development. Meanwhile, two Japanese specialists—agriculturalist Keiji Nishioka and seed-processing technologist Katsuhiko Nishikawa—respectively introduced open pollinated varieties and imported hybrids. The latter sited seed access within commodity relations for the first time. But Bhutan's own National Seed Center has supported a pluralistic approach serving the seed demands of both

M. Kobayashi (✉)
Kyoto University, Kyoto, Japan
e-mail: kobayashi.mai.6t@kyoto-u.ac.jp

Y. Nishikawa and M. Pimbert (eds.), *Seeds for Diversity and Inclusion*,
https://doi.org/10.1007/978-3-030-89405-4_11

family and market-oriented farmers, while organic agriculture became a national mandate in 2007. Bhutan, Kobayashi concludes, has shown that its evolving, idiosyncratic 'middle way' towards food security is likely to endure.

Keywords Bhutan · Indian Green Revolution · Japanese development aid · Seed security

11.1   Introduction and Background

Celebrated for his theories on the origins of Japanese agriculture (Nakao, 1966), the botanist, Sasuke Nakao, was the first Japanese national to visit the Kingdom of Bhutan. In his book *Hikyo Bhutan* [Mystical Bhutan], documenting his five months of travel there in 1958, Nakao recalled a conversation with King Jigme Dorji Wangchuck, the third king of Bhutan.

The Bhutanese king was keenly interested in horticulture, and particularly concerned about the country's future food security, given its growing population. The king opined that importing synthetic chemical fertilizers was probably the best choice, but Nakao told him that they should not run the risk of having to rely on imported inputs and advised instead to plant nitrogen-fixing plants in the rice fields during the off seasons. Nakao promised to send milk-vetch (*Astragalus sinicus*) seeds, which would also be useful as fodder (Nakao, 2013, 151).

We do not know if any milk-vetch seeds were ever sent to Bhutan. However, Nakao's meeting with the king was the beginning of Bhutan's intimate relationship with Japan, which had considerable influence over future narratives of agrarian change. Two years after the exchange, Bhutan's first five-year development plan (FYDP) was drafted, with the support of India, marking a significant step in its emergence from medieval polity to a nation-state. In the plan, Bhutan outlined the establishment of a department of agriculture, which launched a number of model farms and research stations, and set up training for agricultural extension agents in consultation with India, which fully funded all the initiatives (RGoB, 1966a; Savada & Harris, 1993).

Given its steep topography and variable climatic conditions, agriculture in Bhutan remains extremely diverse and employs some 57% of the total labour force (Royal Government of Bhutan [RGoB], 2018). Largely revolving around subsistence-oriented integrated crop-livestock systems, the average landholding is 1.5 hectares, albeit with large regional variation (Renewable Natural Resources Statistics Division, 2019). Expanding urbanization and an increase in fallow land in rural regions has led to a shortage of farm labourers, prompting the government to call for expanded commercialization of the agriculture sector to achieve food and nutritional security (Gross National Happiness Commission, 2018).

Today, Bhutan has come to be best known as a remote kingdom that promotes a unique developmental paradigm, emphasizing happiness and founded on equity and cultural and ecological preservation. But it was far from being independent of global 'modernization' projects, following what Henry Bernstein would describe as a shift from 'farming' to 'agriculture' (2010). This chapter is a brief exploration into Bhutan's modern history, from the last half of the twentieth century to the present, by looking at the interplay of actors within the seed sector. It focuses, in particular, on specific external interventions from India and Japan, and their interaction with endogenous practices, as Bhutan navigates the terms of its autonomous existence.

11.2   A COUNTRY IN SEARCH OF ITS OWN PATH

Bhutan's emergence onto the world stage took place in a precarious era for regional geopolitics. India had just gained independence from the United Kingdom in 1947. Two years later Bhutan, formerly a protectorate of British India, signed a treaty with India confirming its own independence. The treaty, however, also established India as the main sponsor of Bhutan's socioeconomic development.

While Bhutan consulted India as it built its foundations, India was itself negotiating its own terms of state-building with the international community as the Cold War began and US foreign policy started to influence the region's development. Taking advantage of India's initial post-independence agricultural policy goals of food self-sufficiency, US philanthropic foundations sent in scientific advisors and funds to increase cereal production, greatly shaping India's agricultural development narratives (Seshia & Scoones, 2003). The foundations focused not on the generation of new knowledge but on public policy, as well as training Indian agronomists and experts in industrial agriculture (Patel, 2013).

By the time Bhutan was initiating its first FYDP in the 1960s, India was starting to espouse notions of modernization and linear progress, becoming the forerunner in the Green Revolution—the introduction of high-yielding varieties and a range of industrial inputs. G.S. Bhalla and colleagues attribute the success of the new seed-fertilizer agricultural technology in Punjab to the 'large public investment in irrigation and power, scientific research, extension services, roads, markets and other rural infrastructure' during that period (1990). Therefore, by the time of Bhutan's second FYDP, its government similarly included schemes to popularize 'improvements' in tools, fertilizer, seeds, and expanded extension services, with more than a third of its budget going towards rural infrastructure, including roads, water supply and electrification (RGoB, 1966b). A report published by the government detailing the third FYDP, launched in 1972, highlighted efforts to introduce 'modern techniques and practices in agriculture' and the 'regional specialization of crops, provision of improved seeds, implements and fertilizers, [and] introduction of new and improved varieties' (RGoB, 1972). While an emphasis was placed on 'modern methods of farming', it is important to note that discourse was still centred around food self-sufficiency in grains, while also placing importance on the development of cash crops.

Synthetic fertilizers and pesticides were imported from India to spearhead Bhutan's push towards agricultural modernization. While the porous borders between India and Bhutan make it difficult to estimate what agrotechnologies were available when and to whom, it is generally acknowledged that synthetic agricultural chemicals were available in Bhutan from the mid-1960s. Bhutan's national government was put in charge of the procurement and promulgation of these new inputs and technologies, and agricultural extension agents were trained by Indian extension officers. The generation that worked in the ministry of agriculture in Bhutan and gained technical training during this time, accordingly adopted a strong allegiance to modern productivist agricultural methods.

During the following decades, Bhutan's efforts towards food self-sufficiency through agricultural modernization took shape via substantial government subsidies, encouraging the use of inputs and improved seeds. The government provided 'free seeds, free fertilizers, and free pesticides from the 1960s to the mid-80s', according to an officer at the National Plant Protection Center (personal communication, 6 November 2015). To this day, in eastern Bhutan, farmers refer to synthetic fertilizers as *zhungka-ki* in Sharchopikha (commonly spoken in eastern Bhutan), which literally translates to 'government fertilizer'. In western Bhutan, synthetic fertilizers are informally called *jaga lue* in Dzongkha (the national language), which translates to 'Indian fertilizer' (Kobayashi et al., 2015). This clearly reflects the close, ongoing relationship between India and Bhutan, and the central role taken by the state to lay the foundations for Bhutan's agricultural development.

In this way, the initial steps Bhutan took as a modern nation-state directly reflected India's own process of agricultural development during its post-independence period, which was in turn influenced by the global political trends that brought the Green Revolution to India. According to the Bhutanese historian Karma Phuntsho, 'the first step for Bhutan to emerge out of the Indian fold onto the international arena as an independent state' was joining the Colombo Plan in 1962 (Phuntsho, 2013). Established by Britain in 1950, this organization—aimed at socioeconomic development in 27 countries of the Asia–Pacific region—brought Bhutan into contact with other member states to foster cooperative economic development (Savada & Harris, 1993, 334). Through the Colombo Plan, Bhutan revisited Sasuke Nakao's promise to support its quest to define nation-building and achieve the critical mandate of food security under its own terms.

11.3   Diversification of Crop Production

When he returned to Japan in 1958, Nakao arranged for an agricultural specialist to assist in developing Bhutan's agricultural sector. He recommended his student Keiji Nishioka for the job. Nishioka arrived in Bhutan in 1964 as an agricultural specialist under the Colombo Plan.

Nishioka was tasked with assisting the country in developing modern agricultural techniques, diversifying crop production in rural households, and meeting growing demand in urban areas. Nishioka began to grow vegetable and rice varieties from Japan on an experimental farm in Paro, Bhutan. The seeds were then examined and released by Bhutan's National Agriculture Seed and Plant Production Program (NASEPP), which was established in 1984 to produce and supply the domestic need for improved varieties of seed and fruit plants (Tshering & Domang, 2004). At this time, the government allowed only open pollinated (OP) varieties in order to limit their dependence on imported resources, to minimize the economic risk associated with the adoption of new technologies, and to ensure Bhutan's autonomy. Nishioka thus introduced only OP varieties, which can still be found in Bhutan today. Many are recognizable by their Japanese-sounding names (Table 11.1). In 1980,

Nishioka became the first foreign national to be awarded the honorific title *Dasho* by the King of Bhutan and continues to be revered as the father of modern agriculture in the country to this day (Dorji & Penjore, 2011; FAO, 1994).

Despite encouragement to expand and diversify production through extensive government subsidies in chemical fertilizers and seeds (Young, 1991), Bhutan's production of a marketable net surplus of food remained limited. A memoir by Yoshiro Imaeda, a pioneering Tibetologist from Japan, described how upon arriving in 1981 in the capital Thimphu, he was shocked to find not a single market selling fresh produce (Imaeda, 2008). Access to genetic resources remained largely reliant on bartering and gift exchange, despite the government's intentions to expand the subsistence-based agricultural tradition towards a market economy.

While international aid agencies categorized Bhutan as one of the poorest among the least developed countries, World Bank analysts knew that this did not reflect on-the-ground realities (Savada & Harris, 1993) in a primarily subsistence agricultural economy based on bartering. What was calculated as the gross domestic product (GDP) was based on a limited private sector controlled by a small group of people surrounding the royal family or with ties to the government (Savada & Harris, 1993). To encourage further GDP growth, however, the first Companies Act of the Kingdom of Bhutan was adopted in 1989 to separate public and joint-sector enterprises from government regulations (Ramakant & Misra, 1996; Savada & Harris, 1993). Accordingly, NASEPP was privatized in 1995, becoming the Druk Seed Corporation.

Among the sources Nishioka used when ordering seeds from Japan was the Takii Seed Company in Kyoto, where one Katsuhiko Nishikawa was among those taking orders. Nishikawa developed an interest in Bhutan through these exchanges. Upon retirement, he became a senior volunteer with the Japan International Cooperation Agency (JICA) in Bhutan, and was placed in the Druk Seed Corporation, where he served between 2006 and 2008.

When Nishikawa arrived, he was shocked to find that the Druk Seed Corporation sold only OP varieties. He soon learned that hybrid seeds were avoided because they incurred a significantly higher upfront cost. There was also a fear that foreign seed companies would take advantage of smaller nations such as Bhutan, and sell them lower-quality seeds (personal communication K. Nishikawa, 2015). Nishikawa noted other developments, such as the country's high dependence on imported food

Table 11.1 Vegetable varieties introduced to Bhutan by Japan

Vegetable	Variety	Year released	Notes
Pea	Usui	2002	*Usui endo*, a variety said to have been introduced by the United States to the Usui region of Osaka (FoodsLink, 2020)
Carrot	New Khuruda	2006	*New Kuroda*, a hybrid, introduced by Katsuhiko Nishikawa, from Takii Seed Co. Released by Druk Seed Corporation (formerly NASEPP)
Radish	Spring Tokinashi	1990	*Tokinashi-daikon*, a spring variety introduced by Nishioka. Released by NASEPP
	Minowase	1990	*Mino-wase-daikon*, introduced by Nishioka
	Miyashige	1990	*Miyashige-daikon*, introduced by Nishioka. Released by NASEPP
	Shogoem Short	1990	Probably a misspelling of *Shogoin-daikon*. A spherical daikon and well-known heirloom variety from Kyoto. Released by NASEPP
Tomato	Nozomi	1990	*Nozomi*. Source unclear, although Japan's Mayukyo Agricultural Network produces a tomato by this name. Released by NASEPP
Mustard greens	Taisai	1990	*Taisai*, possibly introduced by Nishioka. Released by NASEPP
	Takana	1990	*Takana*, possibly introduced by Nishioka. Released by NASEPP
	Neguna	1990	*Mibuna*, possibly introduced by Nishioka. Released by NASEPP
Bulb onion	Senshu Yellow	1990	*Senshu-tamanegi*, a common winter onion possibly related to the Shenshu variety grown in southern Osaka
Welsh onion	Kujo	1990	*Kujyo-negi*, possibly introduced by Nishioka. Released by NASEPP
Chinese cabbage	Kyoto 1	1990	*Kyoto ichi-goh* (Kyoto number one), introduced by Nishioka. Variety released by the Takii Seed Company
Pumpkin	Tetsu Kabuta	1990	*Tekko-nankin*, introduced by Nishioka. Probably the *Tetsu Kabuto* from the Takii Seed Company

(continued)

Table 11.1 (continued)

Vegetable	Variety	Year released	Notes
Watermelon	Asahi Yamato	1990	Possibly from the original Yamato variety developed pre-war in Nara, Japan, and possibly introduced by Nishioka. Regarding the origin of the name, it is uncertain, although an Asahi variety is bred by Kyoto-based Maru-tane Ltd.

Source Modified from reports published by the Agriculture Research and Extension Division, Department of Agriculture, Bhutan (Ngawang, 2017, 2018)

from India, indicating that Bhutan's domestic market could grow if it increased production. Meanwhile, the domestic seed stock and multiplication technology were not very reliable, revealed by numerous complaints regarding the low germination rates of both domestic seeds and those imported from India.

Nishikawa became convinced that Bhutan should lift its partial ban on hybrid seeds, which, under the 2006 Seed Rules and Regulations of Bhutan, were limited to five ornamental species (Ministry of Agriculture, 2006). He was clearly not alone in this assessment, as he was given immediate permission by the Bhutan Agriculture and Food Regulatory Authority in the Ministry of Agriculture to import hybrid seed samples (Nishikawa, 2015). The breeds he introduced were initially cabbage, cauliflower and broccoli, later expanding to carrot, watermelon and squash (2015). The *New Khuruda* carrot shown in Table 11.1 was one of his introductions. While the tendency towards generalized commodity production does not imply that all aspects of agricultural production are commodified, this was still a revolutionary step: the terms for reproduction were formally outsourced to a foreign company, thus firmly placing access to seeds within commodity relations (Bernstein, 2010).

11.4   Mainstreaming and Commodification

Bhutan's tenth FYDP (2008–2013) called for substantial improvements in the delivery of improved seeds, inputs and technology (Gross National Happiness Commission, 2009a), and mandated the introduction of monoculture (Gross National Happiness Commission, 2009b). The Druk

Seed Corporation's continued dependence on government subsidies to meet the costs of production and distribution (Gross National Happiness Commission, 2009a) led to its reincorporation under the auspices of the government in 2010 and renamed the National Seed Center, or NSC (NSC, 2021). The 2018 edition of Seed Rules and Regulations of Bhutan incorporated a new section on hybrid seeds, stipulating that the country's agricultural department, through the NSC, would be the authorizing body for the import and distribution of these seeds (Department of Agriculture, 2018).

Results from a 2014 household survey conducted with 147 households in three districts in western Bhutan (Gasa, Paro, and Wangdue) assessed how small-scale subsistence farms were acquiring their seeds. It showed that an average of 97% of the households were still saving seeds at some capacity, the most common being vegetable seeds, which were saved by 57–80% of households (Kobayashi et al., 2017). The survey also revealed that for the majority of farming households government distribution of subsidized seeds was a more common method of seed procurement than bartering or purchase.

Perhaps as a testament to government subsidies, a 2016 survey by the Department of Agriculture suggests that only 5% of the population considered itself seed insecure (Department of Agriculture, 2017). Farmers can put in a request for specific varieties to the local agriculture extension agent, who processes the request through the NSC and/or Horticulture Division of the Department of Agriculture. Alternatively, farmers may choose to purchase seeds directly from outlets. Some 75% of the surveyed respondents were already using hybrid seeds or would like to use them, on the basis of their higher quality and greater yield (Kobayashi et al., 2017). Yet not all were convinced. Personal interviews revealed that some farmers preferred OP seeds; one described hybrid cabbage as too large, with fewer opportunities to have multiple harvests over an extended period (Dukpa, 2014). And for the vast majority of farmers, hybrid seeds were unaffordable.

Given the continued mandate to enhance food and nutrition security by fostering a transition from subsistence to commercial agriculture, the NSC tries to balance its priorities by 'following the middle path', as described by the director of the NSC at the time (Dukpa, 2014). On the one hand, hybrid seeds support specialization and efficiency in market-oriented farming operations. On the other hand, OP seeds support the livelihoods of diversified family farmers by minimizing cost and dependence on foreign industries, as well as minimizing the risks associated with genetic uniformity characterized by hybrid seed technology. Attempting a balance and an acceptance of pluralism, the principle of the middle path is critically important to Bhutan's style of governance. Such an approach should not be confused with neutrality, which can be a bureaucratic justification for inaction, avoidance of any action, or consent to the status quo, especially by those who hold more power in society. Instead, it reflects the fact that farmers seek to improve their livelihood and increase overall food provision through diverse means.

Bhutan has attempted to pursue the benefits of market-oriented farming operations, while minimizing dependence on imported inputs—for instance, through its dramatic turn away from agro-chemical inputs as the driving force of agricultural modernization and, as an extension, by adopting organic agriculture as a national mandate in 2007 (Kobayashi et al., 2015; National Environment Commission, 2019). These moves can be understood as deepening commodity relations by further expanding farmers' integration into what is a growing organic market (Bernstein, 2010). However, a commitment to organic agriculture necessitates a deeper dialogue around *processes* of food production, not just the outcome of attaining food security. Reshaping food systems around cultural and ecological sustainability and resilience necessitates a fundamental transformation of Bhutan's agricultural sector.

11.5   Conclusion

Following Nakao's visit to Bhutan in 1958, the country's agrarian landscape was gradually transformed, from the late twentieth through the twenty-first centuries. Changes in policies around the introduction and dissemination of seeds were particularly noteworthy. The self-isolated medieval kingdom opened up to exogenous influences, starting with India's embrace of the Green Revolution followed by Japanese interventions, as Bhutan sought to define its own narrative of development.

If agriculture is a social construction, the introduction of any seed is inherently a reflection of a social relationship that was bred into the seed. The intended relationship, however, is re-shaped by the endogenous cultural and historical context in which it is embedded. Many introduced varieties of OP seed materials have become naturalized within the fabric of Bhutanese society, while hybrid seeds continue to work their way into evolving markets and market agencies. Though still largely government-led, Bhutan's efforts to define their own food and agriculture systems will continue through negotiations with new forces in markets, policies and technologies, while maintaining its idiosyncratic commitment towards a pluralistic coexistence by embracing 'the middle path'.

REFERENCES

Bernstein, H. (2010). *Class dynamics of agrarian change.* Fernwood Pub.; Kumarian Press.

Bhalla, G. S., Chadha, G. K., Kashyap, S. P., & Sharma, R. K. (Eds.). (1990). *Agricultural growth and structural changes in the Punjab economy: An input-output analysis.* International Food Policy Research Institute in collaboration with the Centre for the Study of Regional Development, Jawaharlal Nehru University.

Department of Agriculture. (2017). *Agricultural Statistics 2016.* Ministry of Agriculture & Forests, Royal Government of Bhutan.

Department of Agriculture. (2018). *Seed rules and regulations of Bhutan 2018.* Royal Government of Bhutan. http://www.bafra.gov.bt/wp-content/uploads/2015/06/Seed-Rules-and-Regulations-of-Bhutan-2018.pdf

Doe Doe. (2015, November 6). *National Plant Protection Center, MoAF* [Personal Interview].

Dorji, T. C., & Penjore, D. (2011). *Dasho Keiji Nishioka—A Japanese who lived for Bhutan* (1st ed.). Dorji Penjore.

Dukpa, W. (2014, January 28). *National Seed Centre* [Power Point Presentation].

FAO. (1994). *FAO Seed Review 1989–90.* Food and Agriculture Organization of the United Nations.

FoodsLink. (2020). *Usui-endo.* Encyclopedia of Seasonal Vegetables. https://foodslink.jp/syokuzaihyakka/syun/vegitable/endou-Usui.htm

Gross National Happiness Commission. (2009a). *Tenth five year plan, 2008–2013* (Vol. 2). Gross National Happiness Commission, Royal Govt. of Bhutan.

Gross National Happiness Commission (Ed.). (2009b). *Tenth five year plan, 2008–2013* (Vol. 1). Gross National Happiness Commission, Royal Govt. of Bhutan.

Gross National Happiness Commission. (2018). *Twelfth five year plan: 2018–2023* (Vol. II: Central Plans). Royal Government of Bhutan.

Imaeda, Y. (2008). *Bhutan ni miserarete (Fascinated by Bhutan)* (1st ed.). Iwanami Shoten Publishers.

Kobayashi, M., Chhetri, R., & Fukamachi, K. (2015). Transition of agriculture towards organic farming in Bhutan. *Himalaya Study Monographs, 16,* 66–72.

Kobayashi, M., Chhetri, R., Fukamachi, K., & Shibata, S. (2017). Transitions in seed sovereignty in Western Bhutan. *Journal of Environmental Information Science, 45*(5), 21–30. https://doi.org/10.11492/ceispapersen.45.5.0_21

Ministry of Agriculture. (2006). *Seed Rules and Regulations of Bhutan, Royal Government of Bhutan.*

Nakao, S. (1966). *Saibai Shokubutsu To Nōkō No Kigen.* Iwanami Shoten Publishers.

Nakao, S. (2013). *Hikyo Bhutan (Mystical Bhutan)* (5th ed.). Iwanami Shoten Publishers.

National Environment Commission. (2019). *The middle path -National environment strategy 2020* (p. 164). Royal Government of Bhutan. http://www.nec.gov.bt/wp-content/uploads/2020/07/NES-English_web.pdf

Ngawang. (2017). *Improved crop varieties in Bhutan* (p. 26). Department of Agriculture, Ministry of Agriculture and Forests. http://www.bafra.gov.bt/wp-content/uploads/2020/05/Released-varieties-list_2017.pdf

Ngawang. (2018). *Ineventory of released and de-notified crops in Bhutan (1988–2017).* Department of Agriculture, Ministry of Agriculture and Forests.

Nishikawa, K. (2015, July). 西川克彦 *Former JICA Senior Volunteer at Druk Seed Corporation* [Personal Interview].

NSC. (2021). *Vision & Mission | National Seed Center, DoA, MoAF.* http://www.nsc.gov.bt/?page_id=8

Patel, R. (2013). The long green revolution. *The Journal of Peasant Studies, 40*(1), 1–63.

Phuntsho, K. (2013). *The history of Bhutan.* Random House India.

Ramakant, & Misra, R. C. (1996). *Bhutan: Society and polity.* Indus Publishing.

Renewable Natural Resources Statistics Division. (2019). *RNR Census of Bhutan 2019.* Ministry of Agriculture & Forests, Royal Government of Bhutan.

RGoB. (1966a). *1st five year plan (1961–1966).* Royal Government of Bhutan.

RGoB. (1966b). *2nd five year plan (1967–1971).* Royal Government of Bhutan.

RGoB. (1972). *3rd five year plan (1972–1976).* Royal Government of Bhutan.

Royal Government of Bhutan, T. (2018). *12th five year development plan.*

Savada, A. M. ed., & Harris, G. L. (1993). *Nepal and Bhutan: Country studies* (3rd ed.). Federal Research Division, Library of Congress.

Seshia, S., & Scoones, I. (2003). *Tracing policy connections: The politics of knowledge in the Green Revolution and biotechnology eras in India.* https://opendocs.ids.ac.uk/opendocs/handle/20.500.12413/3984

Tshering, C., & Domang. (2004, March 21). Agricultural Marketing Services-Ministry of Agriculture—The Seed Potato System in Bhutan and the way formard. *Methods for Analysing Market and Market Reforms for High-Value Agriculture.* Jahangirnagar University-IFPRI-ADB workshop, Dhaka.

Young, L. J. (1991). Agricultural changes in Bhutan: Some environmental questions. *The Geographical Journal, 157*(2), 172–178. https://doi.org/10.2307/635274

CHAPTER 12

The Third Way of Seed Governance: The Potential of the Seed Commoning in Japan

Norie Tamura

Abstract As both material entities and "packages" of genetic information, seeds are a common—a co-managed natural resource—in communities and regions all over the world. In this analysis of Japan's national potential for a commons approach to seed sovereignty, Norie Tamura looks through the lens of institutional change. In 2017, the revocation of the Main Crop Seeds Act, a law upholding the state-run seed system, triggered a major backlash in Japanese civil society. Many in the movement called for a revival of state control without fully recognizing the part it plays, for instance, in agricultural industrialization. Yet simultaneously, a range of seed commons exist in Japan at the village and prefectural level, as well as through the network maintained by the Japan Organic Agriculture Organization. For seed commoning to spread and consolidate in Japan, Tamura asserts a need for reintegrating the producer–user divide and fostering open communication between seed and civil sectors.

N. Tamura (✉)
Research Institute for Humanity and Nature, Kyoto, Japan
e-mail: ntamura@chikyu.ac.jp

Y. Nishikawa and M. Pimbert (eds.), *Seeds for Diversity and Inclusion*,
https://doi.org/10.1007/978-3-030-89405-4_12

Keywords Commoning · Community-based seed management · Formal seed system · Japan · Seed commons

12.1   INTRODUCTION

One of the most fundamental aspects of farmers' sovereignty is the freedom to choose which varieties to plant. Farmers determine this based on a comprehensive consideration of farm conditions, local climate and ecosystem, food culture, socioeconomic factors and their own cultivation techniques. However, sovereignty cannot exist without guaranteed access to seeds.

In our globalized era, when multinational corporations are taking control of the world's agricultural supplies, the global peasant movement has repeatedly probed the question of who owns seeds. That concern is in fact part of a larger proposition: who owns nature. The study of the commons, in which researchers continue to explore this question, has also provided an analytical lens on seed sovereignty.

Commons are generally recognized as communally used and managed natural resources. In a 1968 article in the journal *Science*, the US ecologist Garett Hardin first proposed "the tragedy of the commons", arguing that effective resource management can only be provided by the state or through markets, as communal resource management inevitably leads to resource depletion. Hardin's concept has had a great deal of traction. However, successful cases of co-management have been reported from all over the world (Berkes et al., 1989; Feeny et al., 1990). Based on those examples, Elinor Ostrom (1990) identified a set of design principles for the long-term viability of commons.

Ostrom and other scholars in this arena have framed a commons as a "third way" of managing resources that can provide more flexible and efficient governance than states or markets. In the twenty-first century, the field of commons studies has been expanded to non-material resources such as knowledge, information and culture (Bollier, 2014). As neoliberal globalization progresses, multinational corporations attempt to privatize all kinds of information with intellectual property frameworks. We are in the "second enclosure movement" (Boyle, 2003; Evans, 2005). The notion of a new commons has become ever more important.

The multifaceted nature of seeds means that they are treated as commons in two ways: as a material resource, and as a source of genetic information (Sievers-Glotzbach & Christinck, 2020). Seeds are a tangible common-pool resource, and community-based collective management provides flexible conservation, provision and redistribution of them on the ground (Balázs & Aistara, 2018; Coomes et al., 2015; Mazé et al., 2020; Rattunde et al., 2020). On the other hand, seeds are also "packages" of plant genetic information and as such a common property of humanity. There are attempts to create a new seed commons based on the framework of other information commons (Kloppenburg, 2014; Kotschi & Horneburg, 2018; Moeller & Pedersen, 2018), the Open Source Seed Initiative in the United States being a prime exemplar (Kloppenburg, 2014).

Most discussions around seed sovereignty so far have implicitly assumed a configuration in which small farmers confront large agribusiness corporations. However, seed commons is the third way of governance, involving neither state nor market. If this is the case, will the change in state management also stimulate the notion of seed commons? In this chapter, I examine this question by looking at Japan's institutional changes in 2017 as an example.

12.2   Japan's Formal Seed System and Its Turbulent History

The formal seed system in Japan has two strands. One is the registration system for varieties under the Plant Variety Protection and Seed Act (PVPSA), which in principle protects all new plant varieties. The PVPSA is the national law under the International Union for the Protection of New Varieties of Plants (UPOV) Convention, and is part of the international plant variety protection regime. Secondly, there is seed provisioning, further divided into two categories: major agricultural crops including rice, wheat, barley and soybeans, and other plant varieties—mainly vegetables and flowers. For the former, the government controls seed supply based on the Main Crop Seeds Act (MCSA). Hereafter, I use the term "public seed system" to refer to this form of management of major agricultural crop seeds.

The MCSA was enacted in 1952 to promote and produce superior varieties of major agricultural crops in response to post-war food shortages. Under the control and budgetary measures of the national government, prefectures had the primary role of supplying seeds. Their tasks included selecting varieties to promote production in their areas, planning the production of the seeds, examining the quality of the seeds produced and controlling their distribution (Hisano, 2017).

The key staples (rice, wheat, barley and soybeans, as indicated above) have historically been bred by exemplary farmers (*tokunouka* in Japanese) in various parts of Japan. However, since the rediscovery of Mendel's laws in 1900, modern breeding techniques have been mainstreamed, and public institutions have emerged as the main players in breeding (Fujimaki, 2013; Ishizumi, 1968). Particularly after the Second World War, in response to a national demand for increased food production, government-led breeding organizations took the initiative on the development of varieties suitable for various regions throughout the country (Fujimaki, 2013).

In the 1980s, as biotechnology advanced, the private sector gradually grew in capacity (Kashihara et al., 2013). In response, the MCSA was amended in 1986, and the designation of varieties to be recommended for cultivation was extended to privately bred ones. However, those bred by public institutions were still favoured as recommended varieties, making it difficult for private companies to enter the seed market (Hisano, 1999; Kashihara et al., 2013). The MCSA came to be seen as an obstacle to the competitiveness of Japanese agriculture by hindering motivation in the private sector to develop major crop varieties. Therefore, it was revoked in 2017, when Japan decided to join the Trans-Pacific Partnership Agreement.

However, this sudden decision caused a huge backlash in civil society and the MCSA rapidly became a matter of public concern (Hisano, 2017). Figure 12.1 shows the number of newspaper articles on the MCSA, based on the database of *The Asahi Shimbun*, one of Japan's major newspapers. In the period from August 1984 to October 2020, a total of 70 articles reporting on the MCSA emerged, 85% of which were published in 2017 or later. It is clear that the decision to revoke the act has rapidly boosted interest in it, and in the public seed system more generally.

Growing social interest in the MCSA was driven by the public nature of major crop seed supply. Therefore, many of the articles on the act discuss the pros and cons of liberalizing supply. In addition, there were many references to the MCSA in election-related coverage, as since 2017 a number of candidates have pledged to reintroduce it as an issue in national or local elections. Similarly, there were many articles on a local

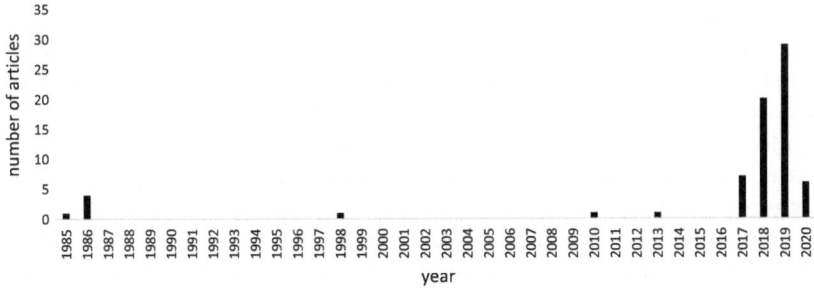

Fig. 12.1   Trends in the number of newspaper articles on the Main Crop Seeds Act

phenomenon emerging after the decision to revoke: to maintain the seed production and supply system, a number of prefectures proactively decided to enact ordinances to replace the MCSA. Civic groups have also gradually set their sights on both reviving the national law and enacting local ordinances. As of October 2020, 22 provinces and prefectures across the country have enacted ordinances that complement the MCSA.

12.3   CRITICAL PERSPECTIVES ON THE PUBLIC SEED SYSTEM FOR MAJOR CROPS

The civic movement regarded the abolition of the MCSA as the dismantling of the public seed system, framing it as a surrender of the public to the private sector by the state (Inyaku & GRAIN, 2020). Members of the movement were concerned that the move would enable an invasion by multinational agribusiness interests. Yet in looking for a means of countering the potential crisis, the movement did not call for seed commons, but rather for a revival of state control. In April 2018, Save Seeds Japan, a leading civic group, released a statement of protest against the repeal of the MCSA. It called for twofold action by the state: to immediately put "legal and budgetary measures" in place "to maintain and develop national and prefectural agricultural experiment stations", and to strengthen the "administrative measures and ordinances of local governments" (Save Seeds Japan, 2018). In my view, this framing of the issues stems from two misunderstandings on the part of the public.

First, there has been insufficient critical examination of public seed systems in Japan. Colin Anderson et al. (2019) propose six critical domains for agricultural transformation: access to natural ecosystems, knowledge and culture, systems of exchange, networks, discourse, and gender and equity. And in fact, the public seed system's evolution in Japan, backed by a series of rice production stabilization policies, had triggered disruptions in these domains.

As mentioned above, public seed systems for major crops were introduced in response to post-war food shortages. Therefore, public institutions initially focused on developing high-yield varieties and technologies. Consequently, rice production had dramatically increased by the 1960s, and self-sufficiency in rice was achieved by 1965. In the 1970s, however, a rice surplus developed, and the government began to reduce production. On the consumption side, consumers' dietary habits and values have become more diversified along with rapid economic growth, and this has made taste and quality central to the value of rice. The rice surplus has also pushed consumption to shape production trends (Nakagahra et al., 1997).

These developments triggered widespread planting of brand-name varieties throughout the country (Kobayashi et al., 2018). Crop diversity was seriously diminished as a result. Rice varieties numbered about 4000 in the late nineteenth century; as of 2005, that number has shrunk to just 88 varieties grown on cropland measuring 500 hectares or more (other varieties are grown on a smaller scale) (Ministry of the Environment, 2011). Prefectural agricultural research institutes began to focus on developing commercially unique varieties, and competition among production areas became more intense (Imabayashi & Yoshida, 1990; Saito, 1990).

To produce quality rice, farmers were instructed to increase the rate of seed renewal. The development of mechanization has separated the seedling stage from the overall production process and barred many farmers from gaining the skills and knowledge involved in seed collection and seedling cultivation. The cost of machine installation also puts pressure on farmers' incomes. Thus, the public seed system has come to function as a part of the modern industrial farming system and has accelerated the commercialization of rice. In this process, traditional agroecological practices have been suppressed and farmers' sovereignty has been weakened. Progressive organic farmers and experienced extension officers have also pointed out these structural dependencies (Hayashi, 1985; Morita, 1994).

Another misconception is that the protest movement has too naive a faith in the state. As James Quilligan (2013) has pointed out, in modern times the word "public" in effect means government, and government is vulnerable to lobbying and other forms of pressure—as he puts it, "captured by elite interests who regularly impede the people's political rights and capacity to control their common goods". Hence, the simple revival of the public seed system will not lead to agroecological transformation. Citizens need to turn their focus to seed commons.

12.4   SEED COMMONS PRACTICES OUTSIDE THE PUBLIC SEED SYSTEM

In the public seed system based on the MCSA, the designation of a variety as desirable has a significant impact on the provisioning of seeds. Prefectural governments take the lead in discussions on designation, and varieties developed by public organizations, both national and prefectural, tend to be given priority. Varieties that have not been designated by the prefecture are difficult to cultivate practically due to inadequate supportive measures, such as difficulties in purchasing seeds and lack of agricultural guidance (Hisano, 1999). And the national government controls the entire system through budgetary measures.

Stefanie Sievers-Glotzbach et al. (2020) define four core criteria for seed commons: collective responsibility, protection from private enclosure, collective and polycentric management, and the sharing of formal and practical knowledge. The public seed system, due to its control by the national and prefectural governments, does not meet these criteria, especially in terms of collective responsibilities and management. Thus it can hardly be regarded as seed commons. For vegetable seeds outside the control of the government, of course, initiatives can be identified that meet these criteria.

The seed network managed by the Japan Organic Agriculture Association or JOAA (discussed in Chapter 7) is a typical example of seed commons. The association, a national organization of organic farmers, considers self-seeding to be an important technical element, and organizes seminars and open meetings to facilitate equal exchange of seeds and knowledge among members. Furthermore, it has established a system to save seeds provided by members and distribute them to those who wish to use them (JOAA, 2002).

The Hiroshima Agricultural Gene Bank is operated by the local government. It not only collects and preserves genetic resources, but also distributes seeds free of charge to farmers within the prefecture. Farmers who take seeds are required to report the results after cultivation, and to collect and return more seeds than they borrowed. Prefectural extension experts provide technical guidance on cultivation and seed collection. Some farmers have revived traditional vegetables that were once cultivated from borrowed seeds, and are now producing them again. Under this scheme, farmers are not just users, but take part in the responsibility of managing the seed bank (Nishikawa, 2001; Nishikawa & Winge, 2013).

In Nagano prefecture, the Seinaiji Akane turnip represents a novel variation on a traditional and local seed saving practice. The turnip is native to the region and has been traditionally served pickled. Around 2000, during a nationwide boom in traditional vegetables, the development of roads in the mountain village area of Seinaiji opened up business opportunities for tourism. The village government and local farmers group decided to commercialize traditional pickles as a specialty product. To do this, they had to standardize the native turnip. Traditionally, however, the vegetable has been maintained in multiple lineages, and traits have varied; and selection of fixed varieties had been attempted before but had not been successful.

The local community therefore collaborated with nearby Shinshu University to develop F1 varieties based on local strains. The developed varieties were jointly registered under the name of the university and the village. Seed production and distribution plans are determined by local farmers' groups. Many farmers in them grow the F1 seeds for commercial ends, as well as the indigenous varieties for their own use (Nemoto, 2012). In Seinaiji village, the community proactively engaged in the development of F1 varieties for the purpose of promoting local agriculture. The F1 varieties bred from local genetic resources can be seen as communally owned new varieties. I see this case as an elaboration of local seed commons practice. The development and registration of F1 varieties based on native varieties can happen not solely as a corporate, private act of enclosure; it can, alternatively, become an endeavour of local commons.

In the case of the Seinaiji Akane turnip, we can also see a local seed commons that has skillfully utilized the plant variety protection scheme based on the PVPSA. However, in 2020, under the influence of the civic seed movement, comments that protested strengthening the protection of breeders' rights through the revision of the PVPSA temporarily went viral on some social networking sites (see Box 12.1).

Box 12.1: A Strategic Approach to Constructive Arguments for Seed Commons

In April 2020, the Japanese actress posted on Twitter that Japanese farmers will be in trouble if the PVPSA is revised to ban self-seeding (Miyahara & Tokubo, 2020). It was the first instance of a celebrity taking up the issue of seeds on social media.

Her post, which warned against the commercialization of seeds, went viral, but her statement was controversial. Her opinion was largely based on the notion of confrontation between big business and small farmers; but the revision in question aimed to protect breeders' rights, and small farmers who grow new varieties can benefit from it. There was an immediate, mixed reaction to her post.

This is an interesting example of pitfalls in the discourse on seeds. Discussion must be open to the public, but public learning on seeds is equally important. There should be a place for repeated discussion and learning, as the scholar Kazuko Tsurumi described: a non-confrontational integrative approach for protesting against governmental and/or corporate power (see also Box 3.1), not just a passing burst of protest.

Reference: Miyahara K., & Tokubo I. (2020, May 21). LDP lawmaker suggests divisive bill to revise Japan plant protection law will be deferred. *Mainichi Daily News*. https://mainichi.jp/english/articles/20200521/p2a/00m/0fp/021000c.

12.5 Envisioning a Future of Seed "Commoning"

That private enclosure is a risk now that the MCSA is revoked is not an unrealistic concern; it may well occur. However, the needed countermeasure is the establishment of seed commons, not a reliance on the conventional public seed system. Seeds as a commons is not an idealist's fantasy. As mentioned in this and other chapters in this book, several different patterns of time-tested seed commons exist throughout Japan.

Recently, a turn to the commons as a social act rather than an analytical tool in academia has been gaining attention (Bollier & Helfrich, 2013; Bollier & Helfrich, 2015; De Angelis, 2019; Euler, 2018). The verbal form of commoning is used in this context to mean regarding certain goods or services actively as commons, and to manage them in a democratic, care-based manner (care here signifying daily action "performed by human beings for their welfare and for the welfare for their community" [D'Alisa et al., 2014]). With seed commoning, the care of seeds is not separated from farmers' everyday lifestyle; it is instead embedded within their life and culture, as are attendant skills and knowledge. Eventually, commoning is expected to provide the foundation for a new socioeconomic system aimed at fostering a sustainable future (Vivero-Pol et al., 2018). Commoning-based socioeconomic structures could emerge from integrating the division between producers and consumers (Varvarousis & Kallis, 2017).

Examples of seed commons in Japan also show that seed producers can be seed users, or that the two groups can be very closely linked. For major crop seeds, reintegrating the division between producers and users may be a starting point for seed commoning. Furthermore, participation is necessary—not only among citizens but also various actors in the seed sector. The civil sector should not only involve in international debates but should also learn about actual practices and struggles on the ground. And the seed sector should be open to civil society. Open communication and mutual understanding between the civil and seed sectors will form a basis for seed commoning in Japan as well as other countries.

References

Anderson, C. R., Janneke, B., Chappell, M. J., Kiss, C., & Pimbert, M. P. (2019). From transition to domains of transformation: Getting to sustainable and just food systems through agroecology. *Sustainability, 11*(19), 5272. https://doi.org/10.3390/su11195272

Balázs, B., & Aistara, G. (2018). The emergence, dynamics and agency of social innovation in seed exchange networks. *The International Journal of Sociology of Agriculture and Food, 24*(3). https://doi.org/10.48416/ijsaf.v24i3.9

Berkes, F., Feeny, D., McCay, B. J., & Acheson, J. M. (1989). The benefits of the commons. *Nature, 340*(6229), 91–93. https://doi.org/10.1038/340091a0

Bollier, D. (2014). *Think like a commoner: A short introduction to the life of the commons.* New Society Publishers.

Bollier, D., & Helfrich, S. (Eds.). (2013). *The wealth of the commons: A world beyond market and state.* Levellers Press.

Bollier, D., & Helfrich, S. (2015). *Patterns of commoning.* Off the Common Books/Levellers Press.

Boyle, J. (2003). The second enclosure movement and the construction of the public domain. *Law and Contemporary Problems, 33*–74. Retrieved February 19, 2021, from https://scholarship.law.duke.edu/lcp/vol66/iss1/2/

Coomes, O. T., McGuire, S. J., Garine, E., Caillon, S., McKey, D., Demeulenaere, E., Jarvis, D., Aistara, G., Barnaud, A., Clouvel, P., Emperaire, L., Louafi, S., Martin, P., Massol, F., Pautasso, M., Violon, C., & Wencélius, J. (2015). Farmer seed networks make a limited contribution to agriculture? Four common misconceptions. *Food Policy, 56*, 41–50. https://doi.org/10.1016/j.foodpol.2015.07.008

D'Alisa, G., Deriu, M., & Demaria, F. (2014). 11. Care. In G. D'Alisa, F. Demaria, & G. Kallis (Eds.), *Degrowth: A vocabulary for a new era*. Routledge.

De Angelis, M. (2019). Commons. In A. Kothari, A. Salleh, A. Esobar, F. Demaria, & A. Acosta (Eds.), *Pluriverse: A post-development dictionary*. New Delhi: Tulika Books.

Euler, J. (2018). Conceptualizing the commons: Moving beyond the goods-based definition by introducing the social practices of commoning as vital determinant. *Ecological Economics, 143*, 10–16. https://doi.org/10.1016/j.ecolecon.2017.06.020

Evans, P. (2005). The new commons vs. the second enclosure movement: Comments on an emerging agenda for development research. *Studies in Comparative International Development, 40*(2), 85–94. https://doi.org/10.1007/BF02686295

Feeny, D., Berkes, F., McCay, B. J., & Acheson, J. M. (1990). The tragedy of the commons: Twenty-two years later. *Human Ecology, 18*(1), 1–19. https://doi.org/10.1007/BF00889070

Fujimaki, H. (2013). *Nihon no ine ikusyu no kiseki to yukue: topics de tsuzuru ine hinsyu kairyo* [Trajectory and future of rice breeding in Japan: Rice breeding in topics]. Record of the lecture at the Japan Agricultural Research Institute.

Hardin, G. (1968). The tragedy of the commons. *Science, 162*(3859), 1243–1248. https://doi.org/10.1126/science.162.3859.1243

Hayashi, N. (1985, February). Kome no hinshu: Hoshii tanemomi ga naze nyuushu dekinai? [Rice varieties: Why can't I get the rice fir I want?]. *Gendai Nogyo*, 248–252.

Hisano, S. (1999). Shuyou nousakumotsu shusi seidoka no kome shushi shijo to agribusiness no jigyou tenkai [The Japanese rice seed market under the 1986 Main Crop Seed Law and agribusiness strategies]. *The Review of the Society of Agricultural Economics, 55*, 73–85. http://hdl.handle.net/2115/11180

Hisano, S. (2017). *Shuyou nousakumotsu shusihou haishi no keii to mondaiten -kouteki syusi jigyou no yakuwari wo aratamete kangaeru-* [The process and problems with the abolishment of the Main Crops Seed Law of Japan: A need to revisit the role of public seed programme]. Kyoto University. http://www.econ.kyoto-u.ac.jp/dp/papers/j-17-001.pdf

Imabayashi, S., & Yoshida, T. (1990). Fukuokakenn ni okeru suitou ikushu no torikumi to kongo no houkousei [Rice breeding in Fukuoka Prefecture]. *Journal of Agricultural Science, 45*(6), 256–258.

Inyaku, T., & GRAIN. (2020). *Trade deals handing Japanese seeds to multinational corporations*. Retrieved July 1, 2021, from https://grain.org/en/art icle/6532-trade-deals-handing-japanese-seeds-to-multinational-corporations

Ishizumi, K. (1968). Wagakuni ni okeru suitou ikushu sosiki no utsurikawari to sono seika [Changes in the organization of rice breeding in Japan and their results]. *Journal of Rural Problems, 4*(1), 1–8. https://doi.org/10.7310/arf e1965.4.1

JOAA. (2002). Shubyou network kiyaku [Seed network rules]. Retrieved July 1, 2021, from https://www.1971joaa.org/

Kashihara, M., Kubo, T., Komura, T., & Komari, T. (2013). Minkan kigyou ni yoru ineikusyu heno tyousen to kongo no kadai [Rice breeding programs of private companies in Japan and future challenges]. *Breeding Research, 15*(4), 173–183. https://doi.org/10.1270/jsbbr.15.173

Kloppenburg, J. (2014). Re-purposing the master's tools: The open source seed initiative and the struggle for seed sovereignty. *The Journal of Peasant Studies, 41*(6), 1225–1246. https://doi.org/10.1080/03066150.2013.875897

Kotschi, J., & Horneburg, B. (2018). The Open Source Seed Licence: A novel approach to safeguarding access to plant germplasm. *PLOS Biology, 16*(10). https://doi.org/e3000023-10.1371/journal.pbio.3000023

Kobayashi, A., Hori, K., Yamamoto, T., & Yano, M. (2018). Koshihikari: A premium short-grain rice cultivar—Its expansion and breeding in Japan. *Rice, 11*(1), 15. https://doi.org/10.1186/s12284-018-0207-4

Mazé, A., Domenech, A. C., & Goldringer, I. (2020). Commoning the seeds: Alternative models of collective action and open innovation within French peasant seed groups for recreating local knowledge commons. *Agriculture and Human Values, 38*, 541–559. https://doi.org/10.1007/s10460-020-10172-z

Ministry of the Environment. (2011). *Annual report on the environment, the sound material-cycle society and biodiversity in Japan.* Retrieved July 1, 2021, from https://www.env.go.jp/en/wpaper/2011/index.html

Moeller, N. I., & Pedersen, M. (2018). *Open source seed networking: Towards a global community of seed commons: A progress report.* HIVOS.

Morita, S. (1994). *Shuyounousakumotsu shushihou haishi heno kougi to nihon no shushi wo mamoru undou sengen* [Protest against the abolition of the major crops seed law and declaration of movement to protect seeds in Japan]. Nosan Gyoson Bunka Kyokai.

Nakagahra, M., Okuno, K., & Vaughan, D. (1997). Rice genetic resources: History, conservation, investigative characterization and use in Japan. In

T. Sasaki & G. Moore (Eds.), *Oryza: From molecule to plant.* Springer Netherlands.

Nemoto, K. (2012). Conservation and management of plant genetic resources in Nagano Prefecture, Japan. *Journal of the Faculty of Agriculture. Shinshu University, 48*(1–2), 85–92. http://hdl.handle.net/10091/15721

Nishikawa, Y. (2001). Chihou hinshu no katsuyou niyoru gene bank to nouka no atarashii kankei: Hiroshimaken gene bank wo jirei toshite [New relations between gene bank and farmers in practical utilization of land-races: A case of Hiroshima Agricultural Gene Bank. *Journal of Agricultural Development Studies, 12*(1), 76–83. Retrieved (abstract only and full text is only in printed version) July 21, 2021, from https://agris.fao.org/agris-search/search.do?rec ordID=JP2002000370

Nishikawa, Y., & Winge, T. (2013). The Hiroshima Agricultural Gene Bank: Re-introducing local varieties, maintaining traditional knowledge. In R. Andersen & T. Winge (Eds.), *Realising farmers' rights to crop genetic resources.* Routledge. https://doi.org/10.4324/9780203078907

Ostrom, E. (1990). *Governing the commons: The evolution of institutions for collective action.* Cambridge University Press.

Quilligan, J. B. (2013). Why distinguish common goods from public goods? The wealth of the commons. In D. Bollier & S. Helfrich (Eds.), *The wealth of the commons: A world beyond market and state.* Amherst, MA: Levellers Press.

Rattunde, F., Weltzien, E. Sidibé, M., Diallo, A., Diallo, B., vom Brocke, K., ... Christinck, A. (2020). Transforming a traditional commons-based seed system through collaborative networks of farmer seed-cooperatives and public breeding programs: The case of sorghum in Mali. *Agriculture and Human Values, 38*(1), 561–578. https://doi.org/10.1007/s10460-020-10170-1

Saito, S. (1990). Meigaramai "Akita Komachi" no ikusei to ryuutsuutaiou [Breeding and marketing of brand rice 'Akitakomachi']. *Journal of Agricultural Science, 45*(1), 14–18.

Save Seeds Japan. (2018). Shuyounousakumotsu shushihou haishi heno kougi to nihon no shushi wo mamoru undou sengen [Protest against the abolition of the major crops seed law and declaration of movement to protect seeds in Japan]. Retrieved July 26, 2021, from https://www.taneomamorukai.com/seimei

Sievers-Glotzbach, S., & Christinck, A. (2020). Introduction to the symposium: Seed as a commons—Exploring innovative concepts and practices of governing seed and varieties. *Agriculture and Human Values, 38,* 499–507. https://doi.org/10.1007/s10460-020-10166-x

Sievers-Glotzbach, S., Tschersich, J., Gmeiner, N., Kliem, L., & Ficiciyan, A. (2020). Diverse Seeds–Shared Practices: Conceptualizing Seed Commons. *International Journal of the Commons, 14*(1), 418–438. https://doi.org/10.5334/ijc.1043

Varvarousis, A., & Kallis, G. (2017). Commoning against the crisis. In M. Castells (Ed.), *Another economy is possible: Culture and economy in a time of crisis*. Polity.

Vivero-Pol, J. L., Ferrando, T., De Schutter, O., & Mattei, U. (Eds.). (2018). *Routledge handbook of food as a commons* (1st ed.). Routledge.

CHAPTER 13

Agroecology, Sovereignty and the Endogenous Development Perspective in Seed Governance and Management

Yoshiaki Nishikawa

Abstract Food sovereignty is a useful concept for researchers, officials and activists eager to understand on-the-ground realities in the context of seed governance. For the Indigenous, peasant and small farmers immersed in those realities, however, it can seem abstract, prescriptive and politicised. In this analysis of the broader debate, Yoshiaki Nishikawa suggests that the Japanese sociologist Kazuko Tsurumi's endogenous development theory offers a perspective applicable to myriad seed-sovereignty approaches. That perspective puts farmers' work experience and cultural values first, recognises and supports a genuinely inclusive seed commons and infuses an understanding of rural realities and spontaneous practices into the study and practice of agroecology. Tsurumi emphasised that transformation is never complete: traditional customs and approaches

Y. Nishikawa (✉)
Ryukoku University, Kyoto, Japan
e-mail: nishikawa@econ.ryukoku.ac.jp

© The Author(s) 2022
Y. Nishikawa and M. Pimbert (eds.), *Seeds for Diversity and Inclusion*,
https://doi.org/10.1007/978-3-030-89405-4_13

189

coexist with modern introductions. Thus, the heterogeneous nature of those who manage seeds, and of the resources themselves, needs to be grasped by scientists and policymakers in this arena. As Nishikawa concludes, it is local decisions, not external ideologies, that must come first in guiding analyses on building better seed management systems.

Keywords Colonization · Endogenous development · Kazuko Tsurumi · Self determination · Spontaneous practice

13.1   INTRODUCTION

In many studies on seed governance, debates hinge on a certain set of polarities: traditional vs. modern, subsistence vs. commercial or local vs. global, for instance. Recently, rights-based approaches to seed governance have emerged as a key area of discussion.

A rights-based approach views human rights and democracy as universal values. It is a powerful way of advocating the value of seed sovereignty for a more sustainable society. Yet it can be a double-edged sword. From the perspective of governance and management, such labelling may override the important values seeds carry for Indigenous and lay people, arising from their commitment to and care of this vital resource based on their own systems and practices. There is plenty of active research on sustainable methods of procuring seeds, especially within political economy and political ecology, with the aim of highlighting the pivotal role of seeds in agroecology (Levidow et al., 2014; Pimbert, 2018a, 2018b; Rosset & Altieri, 2017, also see Chapter 1).

Concepts of food sovereignty and agroecology can be a powerful lens on how interested parties in local contexts can—through commitment, care and adaptive behaviour—foster resilience in seed governance (see also Chapter 3). Food sovereignty can also be a useful tool for third parties (such as researchers, government officials and civil society activists) who need to understand and discuss the activities of those directly involved in seed governance. However, globally, there are many communities where people cherish their own ways of maintaining seed systems that are quite outside Western notions of sovereignty, as described in Chapters 4, 5, and 6. As many Indigenous and peasant movements seeking autonomy advocate, sovereignty needs to be seen from the perspectives of those involved, not those of outsiders (Chambers, 2005; Nishikawa & Hamaguchi, 2018; Scoones, 2015).

13.2   Agroecological Framing of Seed Governance Debates

As agricultural modernisation has advanced, seeds have become objects with economic value, serving as counters in market transactions. Yet at the same time, there has been a gradual rise in global awareness of how farmers contribute to crop diversity.

The unsustainable nature of today's resource-consuming and environmentally destructive food and agriculture systems has been widely recognised. Amid growing interest in establishing sustainable alternatives, agroecology as a concept and field has drawn much attention. The Food and Agriculture Organization of the United Nations (FAO) described ten components as conditions of agroecology: diversity, co-creation and sharing of knowledge, synergies, efficiency to reduce external input, recycling, resilience, human and social values, cultural and food traditions, responsible governance, and a circular and solidarity economy. The organisation notes that the introduction of agroecology is transforming the current system, and also has an affinity with farmer-centred approaches (FAO, 2018). Meanwhile, Peter Rosset and Miguel Altieri (2017), leading advocates of agroecology, view the practice and field as integrating science, agricultural practice and social movements (also see Fig. 13.1).

Colin Anderson and colleagues (2020) identified six 'domains of transformation', or interfaces between the existing food system and agroecological potential: access to natural ecosystems, knowledge and culture, systems of exchange, networks, equity and discourse. They go on to describe conditions in each domain that enable or disable the transition to agroecology as a way of evaluating existing policies, institutions and practices. Cross-sections of these domains and different spatial scales—from household to community, territory, national and international—are proposed as important places for this transformation and the uptake of agroecology as a science, practice and social movement. Anderson and his coauthors conclude that the more different domains overlap,

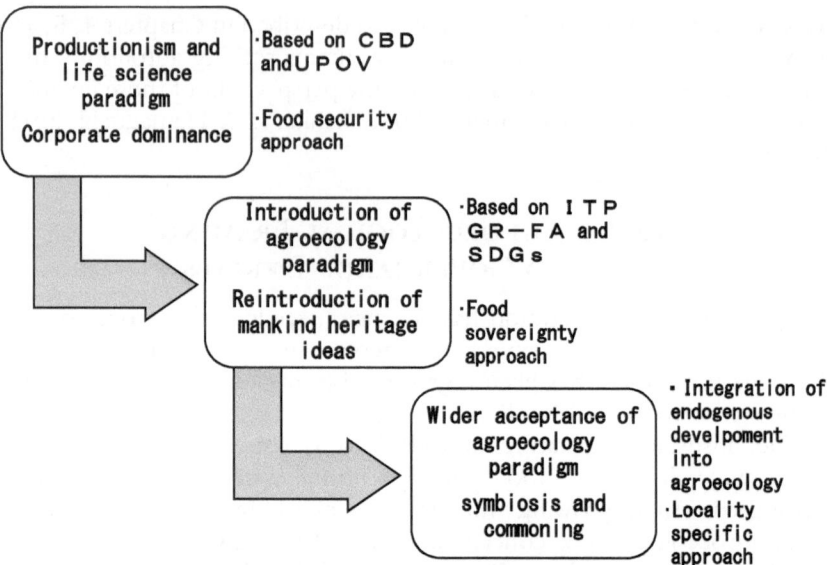

Fig. 13.1 Transformation of seed governance and management; a concept flow (by Author based on Ikegami (2019), Nishikawa (2019))

the greater the potential for durable, pervasive and deep agroecological transformation.

13.3   PUTTING FARMERS FIRST IN SEED GOVERNANCE AND MANAGEMENT

To enrich debates on agroecologically based seed systems, I propose endogenous development theory as a way of describing practices aimed at resilience in Asia and elsewhere (see Chapter 3). The Japanese sociologist Kazuko Tsurumi advocated the endogenous development perspective as key to finding self-determined ways of resource management and development in specific localities, particularly in places touched by environmental disasters such as Minamata, where mercury poisoning from industrial sources led to widespread disease (Tsurumi, 1996; Box 3.1). This perspective, derived from biological concepts of symbiosis, is applicable to a spectrum of approaches to seed sovereignty, such as those seen in this book.

To understand the realities of local people and the rationales behind their processes of knowing and decision making, outsiders need to cultivate a particular outlook: to see not only explicitly observable actions and institutions, but also the process and context behind them (Pimbert 2018b). For instance, Masayoshi Shigeta (1994), over years of research in Ethiopia on the concerns and independence of local farmers, offers perspectives for understanding African agriculture. Before the 1970s, so-called traditional agriculture in Africa, especially mixed cropping, was perceived by Western scholars and development workers as unproductive and inefficient in terms of resource management. Over time, Western observers and theorists gradually realised that such traditional cultivation methods are scientifically sound, as they make effective use of soils and water, help in pest control and contribute to a fair distribution of labour.

Such a shift in perception may be seen as progress, at least in its reevaluation of farmers in Africa. But it is really a cautionary tale in how basing judgement on Western ideas of efficient use of resources (for example) can override the unique experiential wisdom of small farmers. It is necessary to understand the knowledge and ways of knowing that people living in a specific locality have evolved (Sota, 2000; and see Box 13.1).

Box 13.1: Garden Fruit Trees in Japan: Unpaid Labour, Local Values and Maintenance of Diversity

A number of farmers in rural areas of Nagano and Yamagata Prefectures have kept fruit trees in their gardens for many years. The particular municipalities the author visited are famous for their innovative agriculture for small-scale commercial production. Although little economic benefits accrue from these garden orchards, they continue to cultivate them, primarily to provide fruit for family consumption. Typically, they grow persimmon, chestnuts, akebi (chocolate vine), silverberry, plums and pears. Surplus crops are used as gifts exchanged among community residents, especially women.

The cultivation tends to be 'relaxed' and low maintenance, with the work limited to pruning and disinfecting. Despite this, the household heads involved in the practice explicitly recognise the value of their trees as an important ancestral legacy.

> Although all five family heads the author visited appeared not to have strong feelings about uses for the fruits, their wives have continued processing them and preserving the 'taste of family'. While buying agricultural products has become a norm even for farmers, home-processed foods are felt to have a desirable, non-standardised taste.
>
> By maintaining garden orchards, farmers, both household heads and their wives, enjoy non-paid work for 'relaxed care' and 'home processing'. These local spontaneous behaviours, based on their own values and persisting free of external oversight, are prime examples of endogenous development to maintain crop diversity in local areas.
>
> Source: Owada, H. (2019). *Nouka no niwaki kaju no riyou ni miru seizon seikatsuteki na kachi ni kansuru ichi kousatsu [A study on the survival and lifestyle values of farmers' use of fruit trees: From the viewpoint of vernacular and conviviality]* Kyosei Studies, *13(1)*, 98–118.

Agriculture and rural development are applied knowledge, and scientific universalism alone cannot solve or describe their complexities. Farmers' behaviours and decisions depend on local and cultural identities established over time, and based on socioeconomic and political conditions as well as national development policies (Chambers, 2005). Bhutan is a case in point (see Chapter 11). The experience of that nation shows the necessity of a triple approach mingling knowledge of natural science, specific cultural values, and inclusive collaboration in deciding the development strategy of the country.

It is thus key for researchers to understand the seed system from a comprehensive range of viewpoints. If such a deductive flow exists, from the field to the realms of research and policy, agroecology can gain ground in many different societies (Yamane & Ito, 2019).

13.4   THE SEED COMMONS
AND ENDOGENOUS DEVELOPMENT

Tsurumi clearly distinguished between 'bottom-up' and 'top-down' endogenous development (Tsurumi, 1996). In her view, endogenous development should be a social movement in which local residents act or even protest against the central or local governments that promote modernisation policies. However, in many cases, endogenous development can be part of policy, in that central or local governments intend to incorporate into their policies the regional development promoted by local residents who utilise the ecosystem and traditional culture of their community. When concepts to do with agroecology and food sovereignty are imposed on local communities without their consent, the identity and dignity of local stakeholders are eroded.

To avoid this new form of colonisation masquerading as agroecology, many examples gathered in this book offer insights into how to foster diversity and inclusion. They show that diversity can be encouraged in technologies (the integration of hybrid varieties in Bhutan, Chapter 11); networking (cases from Myanmar, Chapter 6; and East Asia, Chapter 7); organisations (small-scale seed companies with family-oriented aspects in Japan, Chapter 10); and strategies (evolutionary breeding, Chapter 8; and seed savers in Japan, Chapter 5).

Individual cases and contexts reveal the overlap between agroecology and endogenous development. Both of these practices also stress the relationship between the environment and the human, and among the environment, science and technology. But there are differences between the two. Endogenous development does not aim to transform the system and the balance of power, which Tsurumi describes as a non-confrontational integrative approach (see Box 3.1). The agroecological aim, meanwhile, is to transform the system through politics as a universally applicable approach, while upholding the imperative to keep within planetary limits.

Shuji Hisano (2017), a researcher and advocate of rights to food, analyses international research trends in the concept of food sovereignty. His work indicates that while food security as a concept is related to normative purpose (a result to be achieved), food sovereignty is related to normative process (paths to be taken or methods to be adopted). An international framework based on rights brings the concept of process into the interpretation of food sovereignty, along with the understanding that agricultural activities are an ongoing reality separate from political economy. Such a framework has the potential to connect agricultural realities with international debates.

Seeds, as biological resources and crop genetic resources, were originally regarded as the common heritage of humankind, well maintained under a diversity of management systems and practices for the commons in specific places. As globalisation and industrialisation have taken hold, commercial control over these resources has grown, to the point where many who are reliant on seeds for livelihoods, identity and tradition have lost the capacity to maintain them sustainably.

As argued in Chapter 12, this situation calls for a re-commoning of seeds. The multifaceted nature of seeds needs to be recognised as a global and local commons, in which the concepts and practices discussed in this section have key roles. The endogenous development perspective can foster the independent, spontaneous development of different management institutions managing local commons, such as the agricultural brokerage guild Poe Yon in Myanmar (Chapter 6), artisanal bakeries in Scotland (Chapter 9) and local seed companies in Japan (Chapter 10). And the concepts and practices of agroecology and food sovereignty can connect these institutions to form global commons, thus transforming an unsustainable seed system to a sustainable one.

13.5   Integrating Agroecology and Food Sovereignty into Seed Governance and Rural Development Realities

Some researchers have established a distinction between agroecology as farming, and as a conceptual framework (Martínez-Torres & Rosset, 2014; Rosset & Martínez-Torres, 2012). They see agroecology as stimulating discourse that in turn inspires people to transform their farming and food systems into sustainable models. By contrast, endogenous development theory asserts that transformative shifts are never complete: customs, languages, and consciousness persist from era to era, and traditional systems and modern systems co-exist like 'uneven rows of icicles hanging down from the remote past to the present' (Tsurumi, 1975).

The idea that agriculture is life-nurturing, and that crops and seeds are entities interdependent with humans, prevails in many cultures and regions. Humans care for and nourish crops and are in turn nourished by them. In some sense, crops and humans become mutually caring partners, as described by the agronomist and smallholder farmer Yutaka Une (2018). If we adhere solely to a political, rights-based approach, where seeds are seen as controllable resources, there is a danger of yielding to neoliberal concepts of seeds' value and relying primarily on modern agrobiology rather than Indigenous knowledge and ways of knowing. We need to look beyond rights as prescriptive and see the practical management of seeds as a process with diverse actors involved, to realise a sustainable society and spark constructive debate.

People whose livelihoods centre on crop diversity procure seeds in truly diverse ways. They do not depend on notions of agroecology and food sovereignty—ready-made ideas brought in from outside. They only want to seek the freedom of self-determination in choosing what to grow and what to eat under the given natural, social, economic and political conditions of their locales. We, as researchers promoting agroecology and food sovereignty, should cherish these spontaneous practices as people's own endogenous ways of exercising normative process—pathways for achieving food sovereignty (see Fig. 13.1).

This book describes the efforts of various actors who manage seeds in diverse natural, social, economic and political contexts, many beset by constraints beyond their control. The heterogeneity of both actors and resources, which are key to the contexts of resource management, need to be recognised by researchers and policymakers (Louafi & Manzella, 2018). From the endogenous development perspective, we need to be careful not to simply apply evaluation criteria set by outsiders in defining what is or is not agroecological and/or endogenous, as discussed in Chapter 3. It is clear that decisions made by locals and communities, not ideologies developed by external interests and actors, are the starting points for analyses of how to build better seed governance and management systems.

References

Anderson, C. R., Bruil, J., Chappell, M. J., Kiss, C., & Pimbert, M. P. (2019). From transition to domains of transformation: Getting to sustainable and just food systems through agroecology. *Sustainability, 11*(19), 5272. https://doi.org/10.3390/su11195272

Chambers, R. (2005). *Ideas for development*. Routledge.

FAO. (2018). *Agroecology knowledge hub*. Retrieved August 6, 2021, from http://www.fao.org/agroecology/overview/overview10elements/en/

Hisano, S. (2017, February 28). Nihon kara miru food sovereignty, food sovereignty kara miru Nihon [Japan from the perspective of food sovereignty, food sovereignty from the perspective of Japan]. Material distributed at the FEAST Food Sovereignty Seminar. Retrieved July 26, 2021, from https://www.researchgate.net/publication/325381570_Japan_from_the_Perspective_of_Food_Sovereignty_Food_Sovereignty_from_the_Perspective_of_Japan_shi liaozhuquankarajianruriben_ribenkarajianrushiliaozhuquan

Ikegami, K. (2019). SDGs jidaino nougyo nouson kenkyu, kaihatsu kyakutaikara hattenn shutai to shiteno nouminzo [A perspective of agricultural and rural studies under the era of SDGs: The necessity of shifting the notion of peasants as project targets to actors in their own development]. *Journal of International Development Studies, 28*(1), 1–18. https://doi.org/10.32204/jids. 28.1_1

Levidow, L., Pimbert, M. P., & Vanloqueren, G. (2014). Agroecological research: Conforming—Or transforming the dominant agro-food regime? *Agroecology and Sustainable Food Systems, 38*(10), 1127–1155. https://doi. org/10.1080/21683565.2014.951459

Louafi, S., & Manzella, D. (2018). The benefit sharing mechanisms under the international treaty; heterogeneity and equity in global resources management. In F. Girard & C. Frison (Eds.), *The commons, plant breeding and agricultural research: Challenges for food security and agrobiodiversity* (pp. 257–271). Earthscan.

Martínez-Torres, M. E., & Rosset, P. (2014). *Diálogo de saberes* in La Vía Campesina: Food sovereignty and agroecology. *Journal of Peasant Studies, 41*(6), 979–997. https://doi.org/10.1080/03066150.2013.872632

Nishikawa, Y. (2019). Jizoku kanouna shushino kanriwo kangaeru [Sustainable management of seeds: Bridging international right based framework and agro-ecology based practices on farm]. *Journal of International Development Studies, 28*(1), 53–69. https://doi.org//10.32204/jids.28.1_53

Nishikawa, Y., & Hamaguchi, M. (2018). Shushi wo meguru shimin soshiki noumin soshiki wo meguru kokusaiteki jokyo ni kansuru kousatsu [A preliminary study on global movement on seeds by civil society and farmers organizations: Perspective through participation in the 7th Governing Body Meeting of the International Treaty on Plant Genetic Resources for Food and Agriculture]. *The Journal of the Society for Studies on Economies and Societies, 58*(3–4), 33–57. Retrieved July 26, 2021, from https://ci.nii.ac.jp/naid/120 006527940

Pimbert, M. P. (Ed.). (2018a). *Food sovereignty, agroecology and biocultural diversity: Constructing and contesting knowledge.* Routledge.

Pimbert, M. P. (2018b). Democratizing knowledge and ways of knowing for food sovereignty, agroecology and biocultural diversity. *Food sovereignty, agroecology and biocultural diversity: Constructing and contesting knowledge* (pp. 259–321). Routledge.

Rosset, P. M., & Altieri, M. A. (2017). *Agroecology: Science and politics.* Fernwood Publishing.

Rosset, P. M., & Martínez-Torres, M. E. (2012). Rural social movements and agroecology: Context, theory, and process. *Ecology and Society, 17*(3), 17. https://doi.org/10.5751/ES-05000-170317

Scoones, I. (2015). *Sustainable livelihoods and rural development*. Agrarian Change & Peasant Studies Series. Fernwood Publishing.

Shigeta, M. (1994). Kagakusha no hakken to nomin no ronri Afurika nogyo no toraekata [Way of knowing by scientists and rationale of farmers, perspectives for African agriculture]. In T. Inoue, O. Soda, & K. Fukui (Eds.), *Bunka no chiheisen* [Perspectives of culture] (pp. 455–474). Sekai-Shiso Sha.

Sota, O. (2000). *Nogaku genron* [Philosophy of agriculture]. Iwanami Shoten.

Tsurumi, K. (1975). Yanagita Kunio's work as a model of endogenous development. In K. Tsurumi, *The adventure of ideas: A collection of essays on patterns of creativity & a theory of endogenous development*. Manga University. Retrieved July 26, 2021, from www.howtodrawmanga.com/products/tsurumi

Tsurumi, K. (1996). Naihatsuteki hatten ron no keifu [Genealogy of endogenous development theory]. In K. Tsurumi & T. Kawata (Eds.), *Naihatsuteki hatten ron [Endogenous development theory]* (pp. 43–64). University of Tokyo Press.

Une, Y. (2018). Noh no soko ni nagareru seishinsei no yutakasa: atarashi nohgaku wo hiraku [Wealth of spirituality flowing in agrarian practices: Towards new agricultural sciences]. *Journal of Organic Agriculture, 10*(1), 36–38. https://doi.org/10.24757/joas.10.1_16

Yamane, Y., & Ito, K. (2019). Datsu kindaika shakai no jitsugen ni muketa nohgaku oyobi nohgyo gijutu shien no arikata [An ideal agricultural science and agricultural technological support for realization of de-modernization society]. *Journal of International Development Studies, 28*(1), 39–52. https://doi.org/10.32204/jids.28.1_39